Wooden Boats *for* Blue Water Sailors

Wooden Boats *for* Blue Water Sailors

Alfie Sanford

GREAT WAVE BOOKS
Nantucket 2021

Copyright © 2021 by Alfred F. Sanford
All rights reserved.

Published by Great Wave Books an imprint of Sanford Boat Co.,Inc.

Printed on demand by Ingram-Sparks in several locations.

First edition

ISBN: 978-0-578-75297-6

Book design by Cecile Kaufman

To Sandi, who assured me I could write this book, well before I knew that I could.

Contents

PART I *Why, How, and What* 1

1. Sea Sailing 3
2. Sympathy with the Sea 27
3. Pioneering Cold-Molding 41

PART II *Science* 65

4. Sailboat Structure 67
5. Why Wood? 86
6. Wood as a Structural Material 95
7. The Stringer-Built Boat 103

PART III *Building with Cold-Molding* 115

8. Lofting 117
9. The Mold 125
10. Building the Hull 136
11. Installing the Interior 156
12. Building the Deck 164
13. At the Rail 173
14. Conclusion 182

AFTERWORD 185

15. Restoring Traditionally Built Craft 187

APPENDIX A Reading the Tape 195

APPENDIX B Drawings and Dimensions 199

BIBLIOGRAPHY 205

ABOUT THE AUTHOR 213

PART ONE

Why, How, and What

As for those who say that men did but use the wind as an instrument for crossing the sea, and that sails were mere machines to them, either they have never sailed or they were quite unworthy of sailing. It is not an accident that the tall ships of every age of varying fashions so arrested human sight and seemed so splendid. The whole of man went into their creation, and they expressed him very well; his cunning, and his mastery, and his adventurous heart. For the wind is in nothing more capitally our friend than in this, that it has been, since men were men, their ally in the seeking of the unknown and in their divine thirst for travel which, in its several aspects—pilgrimage, conquest, discovery, and, in general, enlargement—is one prime way whereby man fills himself with being.

—Hiliare Belloc, *On Sailing the Sea*

CHAPTER 1

Sea Sailing

SAILORS INVENTED BOATS to go places. Long, long ago, far, far away, some caveman fashioned a floating contrivance to sail his lady across the river. Why? To visit her parents, to gather wild berries, just to get to the other side and have a look around? It was risky, but it beat swimming and the accomplishment was satisfying.

Millennia have passed, and boats are still for going places and to get to the other side. After placing their souls at the mercy of Poseidon's vengeance and the caprice of Aeolus' wind bag, sailors still take satisfaction in foreign landfall. Remarkable, indeed, is the experience of casting off lines from your home harbor using this instrument called a sailboat, turning a spoked wheel and pulling ropes, and arriving a few weeks later at a strange port across the sea. There is nothing else like it.

Sea crossing ships take their specific form from what they carry. Freighters carry goods; ferries carry vehicles; tankers carry liquids; liners carry travelers. For the landsman, boats and ships and sailors exist to take him where he wishes to go. They do this on their schedule and at the landsman's expense.

1-1 **NORTHERN LIGHT.** Slocum was part owner and master of NORTHERN LIGHT from 1881 to 1885. She was the finest medium clipper sailing at the time.

For sailors, themselves, there is a special kind of boat, called a yacht, whose cargo is the sailor himself. She allows him to sail her where he wants, on his own schedule and, of course, at his own expense. The sailor leaves home aboard his yacht, to make a voyage of personal discovery. Living at sea, dependent on his own resources, he sails into a transcendental wilderness untrammeled by society. He sails seeking knowledge of himself that he may begin to understand the God in whose image he is made. From his voyage he learns about the home he left behind. His is a noble search. This book is about his accessory in that search—his boat—and how to build her.

Ocean sailing, as personal venture, began in the late 19th century. Early blue water yachtsmen used craft based on the work boats that fishermen and pilots had developed over the centuries. In 1895 Joshua Slocum established the pattern by rebuilding, with his own hands, a derelict fish boat and sailing her around the world.

Slocum had been a master of sail whose profession was eliminated as steam power killed off the great age of sail. Unlike a Magellan or Cook, whose voyages were made by royal command on the king's ships, Slocum set off around the world on his own boat and at his own command.

1-2 An abandoned oyster boat, *SPRAY* was gifted to Slocum in 1891. He rebuilt her and sailed her around the world 1894 to 1898. *SPRAY* was noted for her extraordinary self-steering qualities, qualities that have rarely been equaled and never scientifically explained.

PRESTO — Ralph Munroe

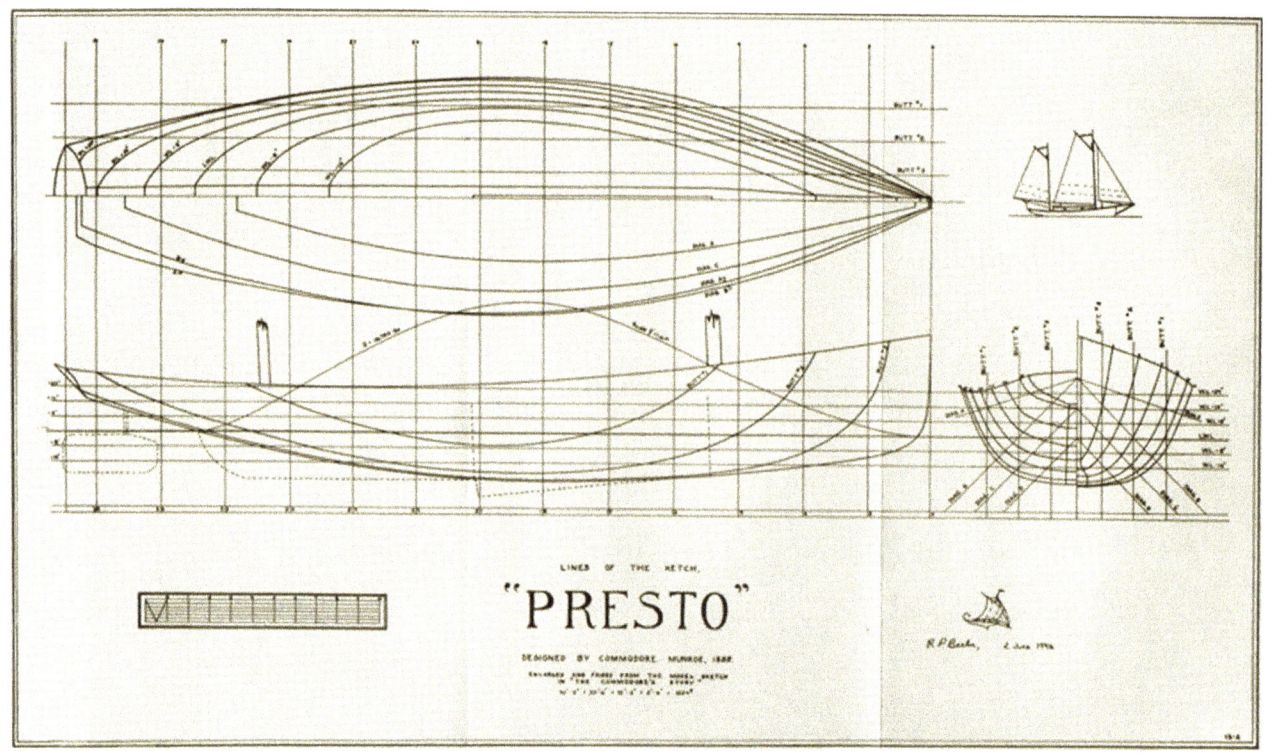

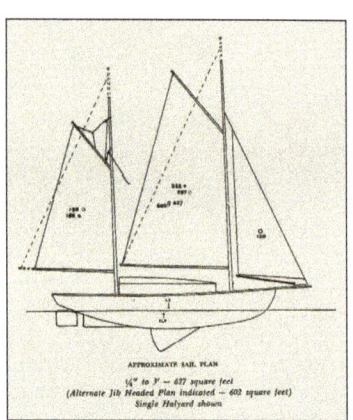

Created by Ralph Munroe in 1885, *PRESTO* predates *SPRAY*, or at least predates *SPRAY*'s time as a yacht. It is a proper question to ask, was she Munroe's yacht, or was she his age-of-sail workboat, his ocean going pick-up truck? Slocum had given up the sea as a profession when he began sailing *SPRAY*. Munroe, a waterman from Long Island, sailed *PRESTO* to Florida because that was the practical way to get there. Once there, he used her to deliver the US mail to Florida coastal settlements. Before Flagler's railroad, boat was the only way to travel Florida—the impenetrable Everglades prevented inland travel, and while one could ride a horse locally up and down the beach, every ten miles or so an inlet stopped the horse. Munroe's homestead, "Coconut Grove," on Biscayne Bay was the beginning of what is now Miami.

PRESTO was an extremely shoal-draft, 41 foot, round bottom variant of the New Haven sharpie. Munroe made regular winter offshore passages from Long Island to Florida with her. Although he did not really use her as a yacht, Munroe used her model in later yachts he designed for others, including *ALICE* (see adjacent). Most of Munroe's work was lost in a 1926 hurricane, but Vince Gilpin's book, *The Good Little Ship*, describes her. *The Commodore's Story*, Munroe's biography, describes the world she came from.

ALICE — Henry Howard

1-4

Ralph Munroe designed *ALICE* in 1921 for Henry Howard and his wife Katherine, extending his *PRESTO* concept to a larger boat. This is a little new, for here we have a professionally designed boat built for an amateur yachtsman who will use it, part-time, for vacations from city life. Blue water sailing is becoming a popular pastime in the 20's. *ALICE* is shoal draft and set up for shorthanded sailing. A great deal of attention has been paid to the comfort of her crew. The Howards cruised her extensively for over 20 years and wrote two books about her, first *The Yacht "ALICE,"* 1926, and then *The Yacht "ALICE" Twenty Years After,* 1946.

ISLANDER

ISLANDER begins to take shape.

Launching the new vessel in Los Angeles harbor.

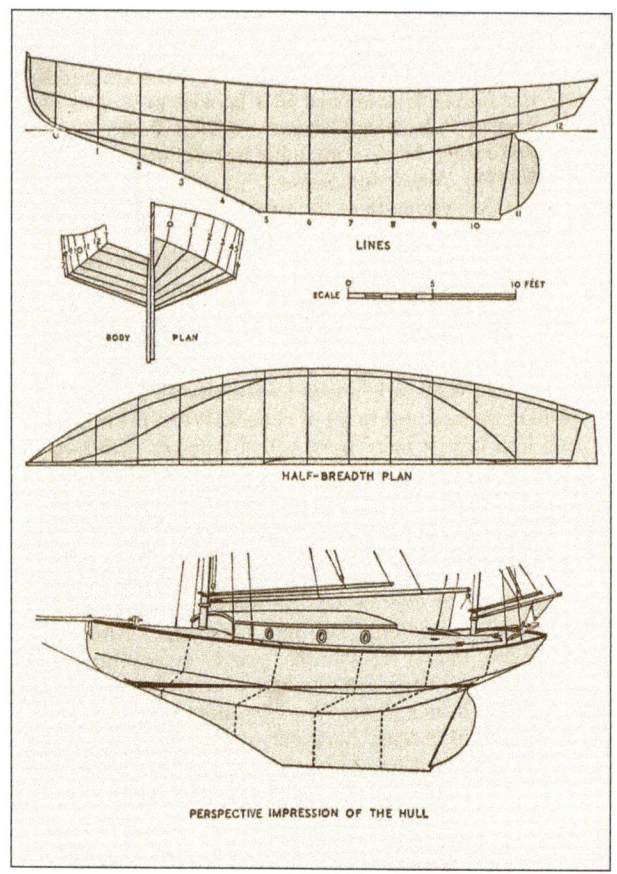

1-6 *ISLANDER* careened.

Harry Pidgeon, an inland man from Iowa, began his voyaging on rivers, first the Yukon, then the Mississippi. He developed a yearning for the deep sea and in 1916 discovered Thomas Fleming Day's 35-foot *Sea Bird Yawl* design in *Rudder Magazine*. Tom Day not only started the Bermuda Race, he started *Rudder Magazine*. He inspired Pidgeon and many others to go to sea in small boats. Pidgeon built ISLANDER by himself on the beach at Los Angeles when he was 48 years old. He used $1,000 and 18 months of his time. ISLANDER had no engine.

Many *Sea Birds* were built over the years. They came in two sizes: Pidgeon's 35-footer and a 25-foot version, which John Voss, another early voyager, sailed through a typhoon in Japanese waters in August of 1912. In the late 1930's, Frank Wightman built his 35-footer, *WYLO*, again on the beach and instructed by Pidgeon's example, this time at Capetown.

After a shake down to Hawaii, Pidgeon sailed *ISLANDER* around the world, 1921–1925. Pidgeon was self-reliant. Not only did he build his own boat, he took care of her as well without cranes or boatyards. Pictured here, she is careened at Nuku Hiva for a bottom job.

Pidgeon sailed *ISLANDER* around the world again in 1932. Then, at age 72, he married (for the first time) and set out on a third circumnavigation, which came to ruin in the New Hebrides. *ISLANDER* was wrecked at her age 31, Pidgeon's age 79. He died in Los Angeles 6 years later.

FORE AN' AFT — William Atkin

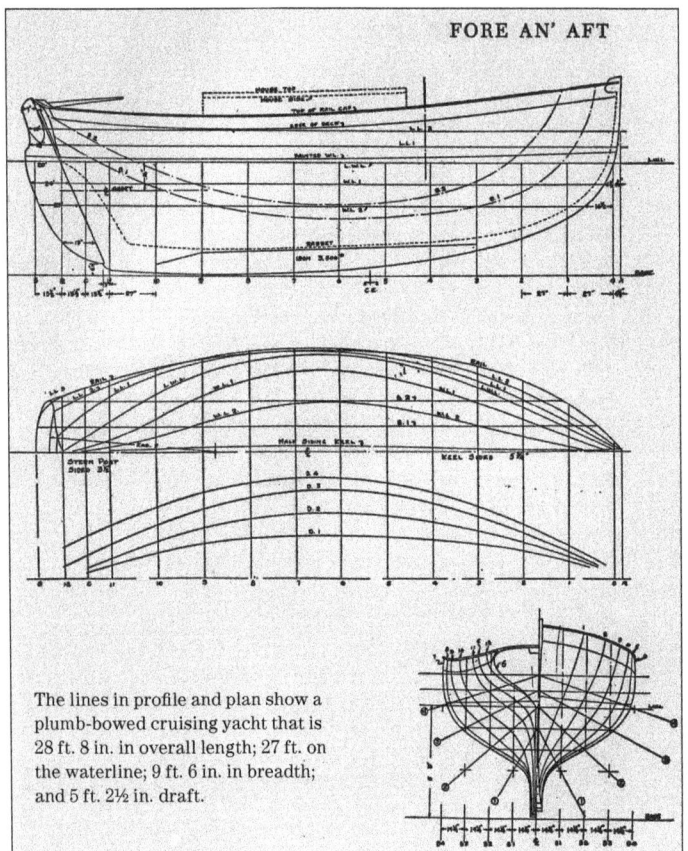

The lines in profile and plan show a plumb-bowed cruising yacht that is 28 ft. 8 in. in overall length; 27 ft. on the waterline; 9 ft. 6 in. in breadth; and 5 ft. 2½ in. draft.

A pen and ink study made by C. G. Davis of the cutter *Fore an' Aft* under full sail

William Atkin began, in the 1920's, designing and developing small cruising yachts for amateur recreational sport. *FORE AN' AFT*, 1927, is a nice example of his work. Atkin began with research into historic working craft, in this case Norwegian Colin Archer's rescue boats. He developed them through his own building and sailing. He publicized his work in the burgeoning yachting press that was led by Tom Day's *Rudder* and a little later by *Yachting Magazine*.

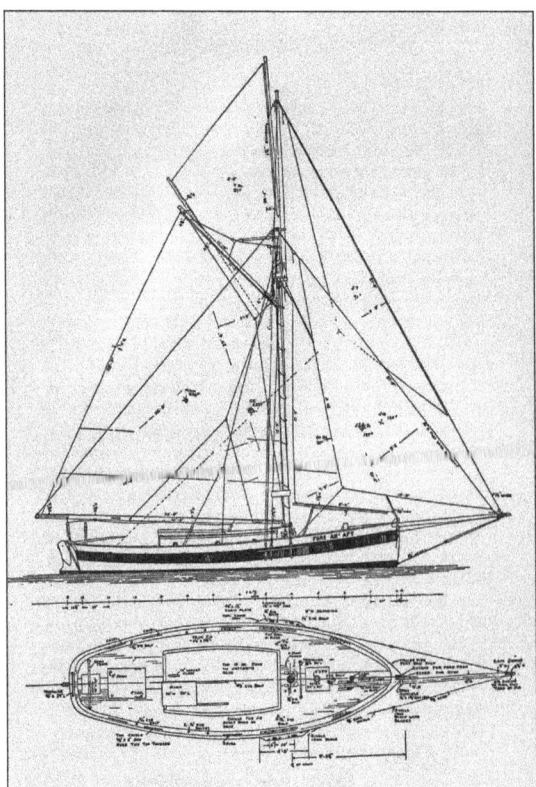

Fore an' Aft has a sail area of 701 sq. ft. distributed as follows: mainsail, 336 sq. ft.; staysail, 72 sq. ft.; jib, 125 sq. ft.; topsail, 97 sq. ft.; and jib-topsail, 71 sq. ft.

SVAAP — William A. Robinson

1-8

   Robinson sailed *SVAAP*, a 32-foot Alden ketch, around the world in 1928–1932 and wrote a good book about the voyage, *Deep Water and Shoal*. Most of the way he sailed with a paid hand! In 1933–1934 he took *SVAAP* to the Galapagos. That trip was terminated by his appendicitis. Robinson was miraculously rescued by the US Navy and *SVAAP* was lost to the Ecuadorian authorities.

   Robinson went on to build his dream boat, 70 foot brigantine *VARUA* of which more later.

MALABAR X — John G. Alden

1-9

Racing brought serious money and attention to the development of the blue water sailing cruiser. This development resulted in elegant creations like the one pictured here. John Alden made a financially successful career as a designer of ocean racing cruisers. Each year he designed and built one for himself, which he would race to advertise his prowess to the public. He reached his peak with his tenth *Malabar*, in 1930, with which he won the Bermuda Race. Alden's *Malabars* were based on traditional fishing boats (*MALABAR X* is a miniature Grand Banks fishing schooner) but aimed at the wealthier yachting market.

MALABAR X, built in 1930, epitomizes the longevity of many beautiful sailing cruisers. I first met her in 1954 as a charter boat in the Bahamas. Next, I saw her again in Barcelona, Spain, in 2011. She was looking just as good as in '54. At that point she was 81 years old, still sailing, still loved.

LEGH II — Vito Dumas

1-10

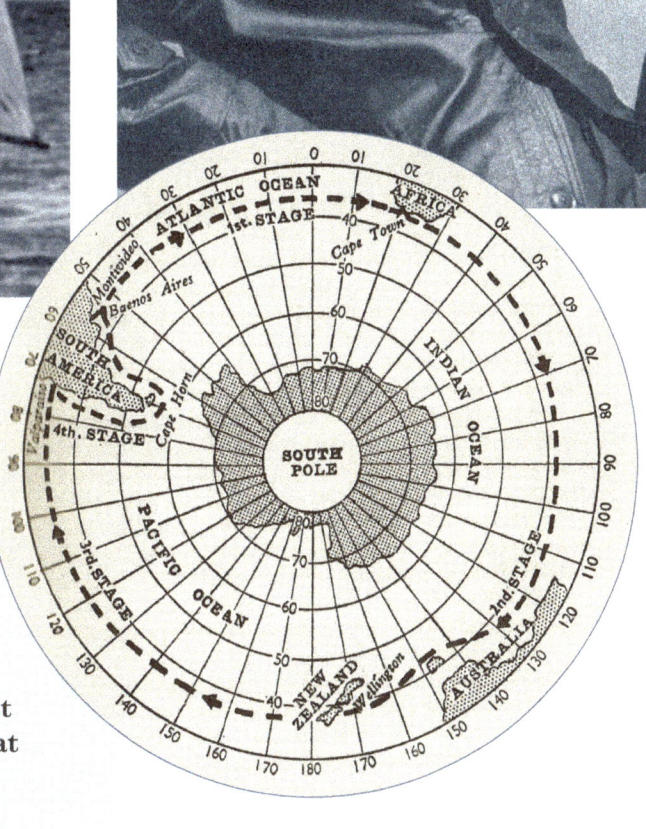

Vito Dumas left Buenos Aires in July of 1942 with the southern world in winter and the northern world at war. Taking about a year with stops at Capetown, Wellington, and Valparaiso, he sailed right around the world, following the parallel 40° South. *LEGH II* was designed and built in 1933 by Manuel Campos. She was a 31-foot Colin Archer style double-ender, a model that both he and William Atkins had become expert with.

Dumas wrote of his sail in *Alone through the Roaring Forties*. He established four firsts with his voyage, but that was not his purpose. He was led by his love for the sea and pushed by his fear of being "one of those creatures chained to the treadmill of today and tomorrow." His was indeed a "magnificent feat"; he was a quiet and great man.

Map of route taken by Vito Dumas in his single-handed voyage around the world.

VARUA   William A. Robinson

VARUA was Robinson's dream boat, conceived and built by him after many years of ocean sailing. She was his idea of the perfect boat to accomplish the voyages that he had begun with *SVAAP*. She was designed by Starling Burgess to be lived aboard and sailed across blue water by Robinson aided by a small crew. Robinson took her to the southern ocean to test her ability in the "ultimate" storm. She passed.

LONE GULL — Maurice Griffiths

Meanwhile, parallel development of small cruisers for amateur yachtsmen was occurring in English waters. Maurice Griffiths, editor of *Yachting Monthly* from 1926 to 1967, was a designer and sailor of small craft designed for cruising Britian's dangerous and interesting coast. *LONE GULL* is his dream ship.

TREKKA — John Guzzwell

1-13

TREKKA is the latest of an old tradition. Although she was professionally designed by Laurent Giles for Guzzwell's 'round-the-world cruise, she was home built by Guzzwell for his voyage. She is distinctive in two ways. First, she is very small—only 20 feet long. And second, she has what, at first glance, looks like a dinghy style hull. But actually it is not. Unlike AVENGER or the *Flying Fifteen*, she is heavy displacement—she had to be to carry John and his cargo of long-distance supplies—and she has a long keel. Like his predecessors, Guzzwell made a good voyage and wrote a good book about it. TREKKA is today on display at the Maritime Museum of British Columbia in Victoria, BC, along with an earlier ocean cruiser from Victoria, the canoe TILIKUM, in which John Voss cruised across the Pacific in 1902.

CHAPTER 1 ∞ SEA SAILING

He was not the first to sail long-distance, or even single-handed; there were many others. The well-known ones were showmen who sailed (or rowed) their craft to professionally entertain landsmen with their phenomenal voyages. Slocum made his voyage for personal reasons. He was at a hard time in his life. Not knowing what to do or where to turn, he did what he knew: he started sailing his smack across the oceans of the world. He came home a better man. Slocum was also a raconteur. With his book, *Sailing Alone around the World*, he told his tale so well that he inspired an age of individuals sailing their own small craft on their own voyages of private discovery.

Slocum's voyage established the requisites for the blue water sailor. Required was the desire to adventure forth, the courage to begin, and the self-reliant skill to accomplish. His oyster smack *SPRAY* embodied the requisites for the blue water sailing boat. She ought to be easy to sail, comfortable to live aboard at sea, and safe in storms—a vessel both seaworthy and sea-kindly. And for success, she wants beauty. For beauty inspires; without it the sailor's quest is threatened by despair.

The design of the blue water cruiser is a job for the artist; her construction will be the artist's conception executed in substance. She must meet the demands of the sea while fulfilling the desires of her crew. She is a multi-ton sculpture incorporating the subtle physical requirements for good performance, ease of handling, and comfort. She materializes the interrelated and contradictory demands of aero-hydrodynamics to get her moving, structure to keep her afloat, and domesticity to care for her crew. Her design is guided by science and measured by the sea. She is built with craftsmanship honoring centuries of experience. Well done, she is transcendent—a thing of beauty that delights the sailor and beguiles the landsman.

In time, many followed Slocum. Some, Harry Pidgeon, John Guzzwell, Larry Pardey, also built the boats in which they sailed. Others used older boats for their voyages. A few had new boats built for their special purpose. Many wrote books developing the design and gear of the blue water sailboat. They were, from America, William Robinson, Irving Johnson, Alf Loomis, Carleton Mitchell; from Great Britain, Hilaire Belloc, Connor O'Brien, Erskine Childers, Claude Worth, Uffa Fox, Ernle Bradford, Donald Street; from France, Alain Gerbault, Bernard Moitessier, Eric Taberly; from Argentina, Vito Dumas. The lists are not complete.

The adventures and books of these great sailors informed two generations of yacht designers—Americans Ralph Munroe, Starling Burgess, Francis Herreshoff, John Alden, the Atkins, Olin Stephens, Aage Neilson, and Phil Rhodes and Europeans Laurent Giles, Maurice Griffiths, Harrison Butler, and Albert Strange. These designers filled the world with gorgeous,

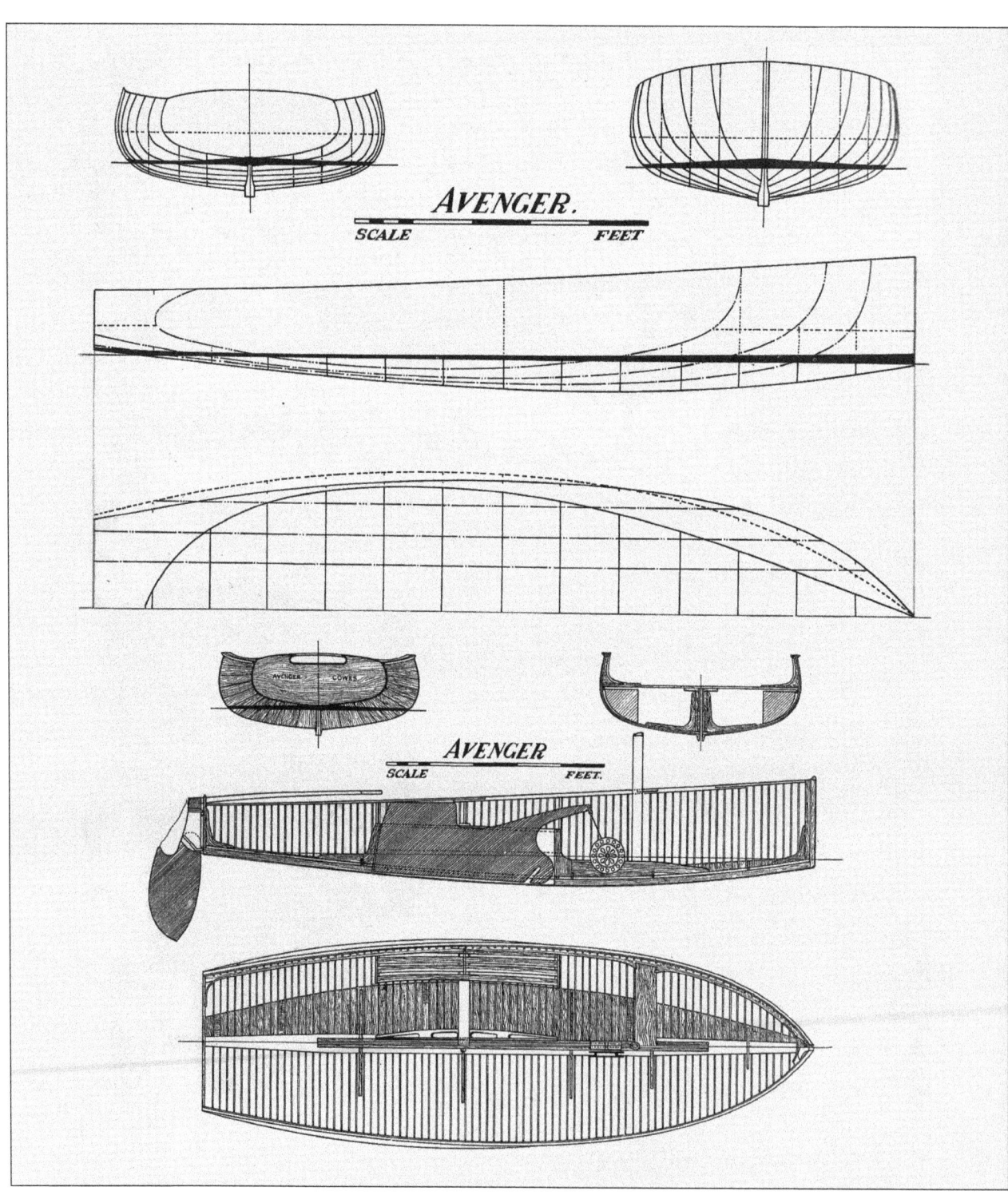

1-14 *AVENGER*. The international 14 footer is Uffa Fox's breakthrough sailboat, built in 1927. He gave her the flat bottom and hard bilges that make planing possible. She was unique in 1927. Her pioneering shape brought her 52 firsts, 2 seconds, and 3 thirds in her 57 starts in the 1928 racing season. She began the age of the planing dinghy.

seaworthy, sea-kindly small vessels for amateur sailors to cruise the world.

As the sport developed, people being people and particularly sailors being sailors, the owners of these craft began to race them. Some of the racing became organized. In 1908, Thomas Fleming Day, who had founded *Rudder Magazine* twenty years before, wishing to demonstrate the seaworthiness of small cruising boats, organized a race from New England to Bermuda. The event was quite controversial at the time because the yachting public did not accept that small boats could safely sail offshore. Day's race proved otherwise and it grew influential.

A race event, like many human competitions, develops over time in a pattern that is timeless. It begins with pioneers. A few brave and brilliant pioneers establish an idea using simple, maybe even obsolete, equipment. The pioneers inspire followers whose money and energy create a flourishing golden age in which the idea and the equipment for pursuing it are perfected. During the golden age the event becomes glamorous, and the glamour draws in a different sort of competitor, a competitor for whom "winning" is the entire purpose. The purpose of the pioneers who brought about the golden age is forgotten, and the golden age degenerates into decadence, with mediocre men aping without understanding, parodying the earlier heroes. The honor and glory move away like Carl Sandburg's fog, "on little cat feet."

This has happened with the America's Cup, which started in 1851 and reached its golden age in the 1930's. It has been all downhill since then, slowly at first, then fast. It happened with long-distance, single-handed racing. The *OSTAR* only survived two races before professionalism obliterated the challenge. It even happens locally amongst work boats. Howard Chapelle chronicles how racing destroyed the development of the Cape Cod catboat in the 19th century,[1] evolving a handy fish boat into a fast and cranky one.

And it happened with the Bermuda Race. The Bermuda Race brought big money into the development of the sea-kindly, seaworthy blue water sailboat. The early boats, a hodgepodge derived from 19th century workboats, evolved into magnificent seagoing yachts, of which the *MALABAR X* is the epitome. After the Second World War, winning became more, but not all, important. Philip Rhodes and Olin Stephens developed the ocean racing yawl, which was a compromise between the speed of International Rule racing boats and the sea-kindliness of the fishing schooner style so popular in the 1920's and '30's. Faster, but less comfortable, the new boats were still seaworthy and sea-kindly enough that their owners used them as summer homes for their families. The Bermuda Race reached its golden age peak when little

---

1. Chapelle, *American Small Sailing Craft*, p. 255.

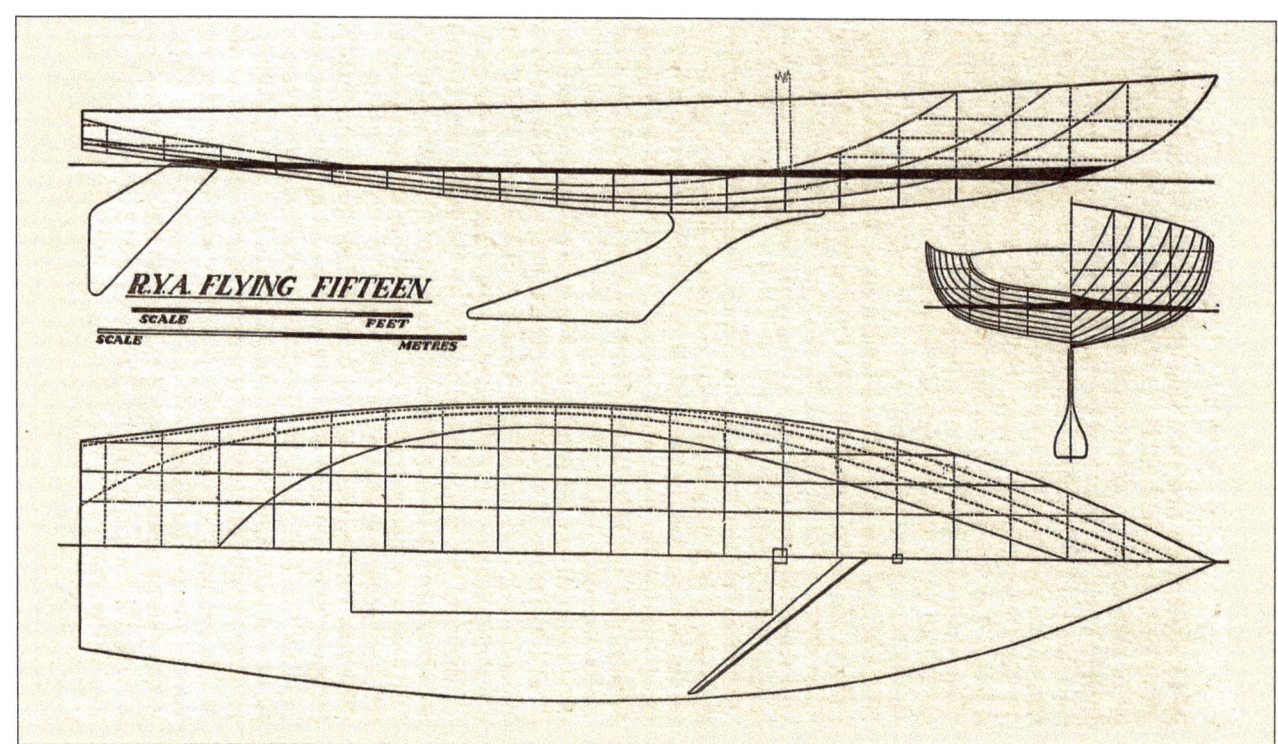

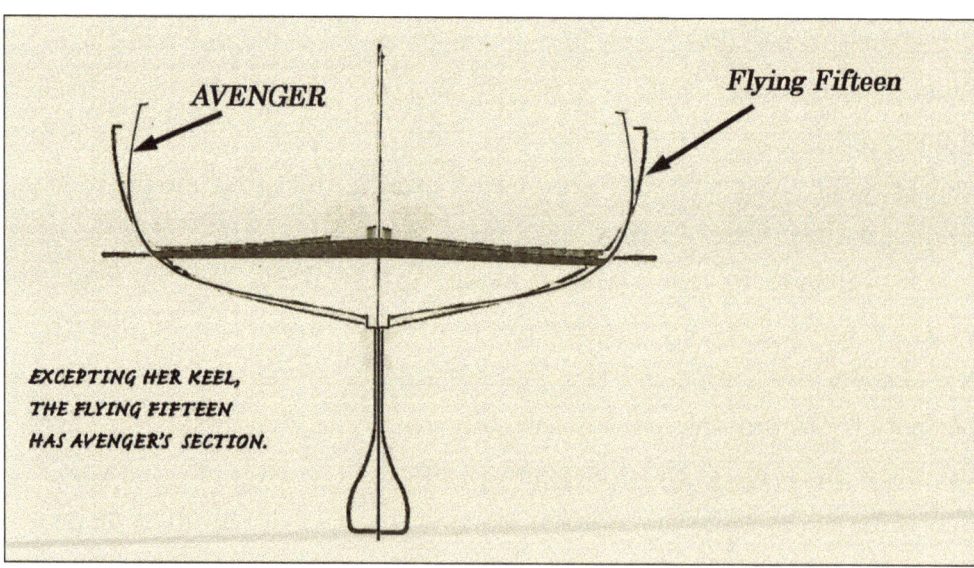

1-15 *Flying Fifteen*. Twenty-five years after *AVENGER*, in 1952, Uffa added a keel to the planing dinghy hull and gave the world the *Flying Fifteen*, the first planing keelboat. Her model is still with us for high-speed monohulls, being surpassed, today, only by multihulls and hydrofoils.

FINISTERRE, Carleton Mitchell's little "ocean-going gunk-holer" won three times in a row, 1956, 1958, and 1960, probably by understanding the Gulf Stream's meanders.

Many innovations pioneered by Bermuda racers carried over into offshore cruisers, developed positively from Slocum's time till about 1965. Then, in a complex way, the mood changed. The offshore cruiser's development took an abrupt turn as two revolutions, one in design and one in construction, swept the sailing world.

The design revolution started 40 years earlier when, in 1927, England's irrepressible Uffa Fox upstarted sailboat racing with his invention of the planing dinghy, *AVENGER*. In 1952 he extended the planing concept to keel boats with the *Flying Fifteen*. The planing boat ignored the demands of seaworthiness and radically increased the upper limit of speed under sail. Unbeatable on protected water, Uffa's conception won all races. While no one will call planing dinghies comfortable, for short spells they are great fun and good sport. They are also ideal for racing—complicated, uncomfortable, demanding, and fast.

In racing, speed trumps seaworthiness, and, generally speaking, speed and seamanship are inversely related. Any seamanship-like development that increases speed soon becomes standard seamanship. This leaves the rest of speed-enhancing developments in opposition to seamanship. Comfort at sea, or sea-kindliness, is an example. What racing crew will eschew a speed increase just because it makes the boat uncomfortable? For them the finish line means hotel and shipyard, the race is over, and any reserve of stamina or vessel at the finish is by definition an unnecessary expense on speed.[2]

Through the 1950's and 1960's the ocean racing fraternity, under the stern guidance of the Cruising Club of America (CCA) and the Royal Ocean Racing Club (RORC) in England, disregarded Uffa's inventions and the newly attainable sailing speeds. Traditional deference to seamanship kept dinghy hulls out of the offshore fleets, *HOOT MON*,[3] an early and scandalous exception, notwithstanding.

Then in the mid 1960's, speed-envy overcame good sense. Dick Carter was the catalyst for change. He was an International 14 sailor who had ventured into ocean racing with his first *RABBIT*, a *Medalist Class* 32-footer designed by Bill Tripp and built by Dolf LeCompte in Jutphass, Holland. She was the latest thinking in CCA-inspired design and built in the new material fiberglass. (Carter's seductive advertisements for LeCompte, featuring *RABBIT*, prompted my purchase of one, *WIND*, discussed in chapter 2.)

---

2. Dick Nye said of his *CARINA*, a 53 foot Rhodes centerboard yawl, as she crossed the finish line at the end of the rough 1957 Fastnet Race with three broken frames, a deck that was threatening to part company from the hull, and most of the crew manning the pumps, "Okay, boys, you can let her sink."

3. *Faster Sailing*, p. 45.

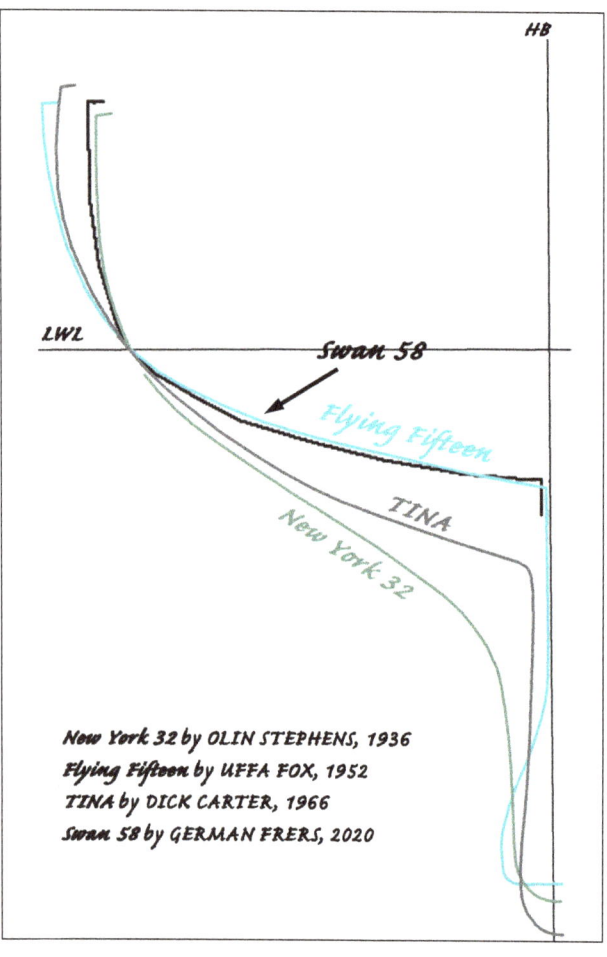

1-16 Underbody and speed.

Carter, starting from his experience with the *Medalist*, came up with his own design, again called *RABBIT*, and won the 1965 Fastnet Race. *RABBIT* was a good-looking boat. She was radical in purpose—she was built to win—but only cautiously radical in design. But with her trophies on his shelf, Carter then designed with less caution. *TINA* and *RED ROOSTER* forever changed ocean racing and ushered in the International Offshore Rule (IOR) in 1970. At this point ocean racing and ocean cruising parted ways. Carter tells his tale in his recent book, *Dick Carter, Yacht Designer*.

The IOR, pushed strongly by the Europeans, who were more interested in racing their boats than going places with them, encouraged dinghy style competitors. Designs developed from the models of *AVENGER* and the *Flying Fifteen* took over the winner's circle; the traditional designs became obsolete. The era of the sea-kindly ocean racer was over.

This illustration compares the shapes of three hulls. The boats are of different sizes and the drawings have been adjusted to make waterline beam the same for all three. There are two shapes that allow a monohull to achieve high speed. A very narrow hull will cut through the water reducing the normal speed-limiting wave train; a very light hull will skim across the top of the water, or plane, leaving its wave train behind. For sailboats both shapes have disadvantages. The narrow sailboat lacks the stability to carry sail. The planing sailboat has a flat bottom that pounds in a seaway. The mid-section of the *New York 32* shows the traditional compromise that moves through a seaway smoothly, but at a speed limited by its wave train. *TINA* is much lighter (the area of the midsection, under the waterline, is a proxy for weight). She will be faster at the expense of pounding. The *Flying Fifteen* will actually lift out of the water and plane and has a theoretical speed limit of several times that of the *New York 32*. Today the *Flying Fifteen* shape has come to dominate ocean racing. Frer's *Swan 58* (see figure 1-18) uses it exactly.

1-17 *HOOT MON*. Built in Miami in 1952, *HOOT MON* was not a particularly successful attempt to use light displacement to win ocean races. Despite her failure, racers did not give up their effort to dominate ocean racing by means of high-speed, light-displacement hulls.

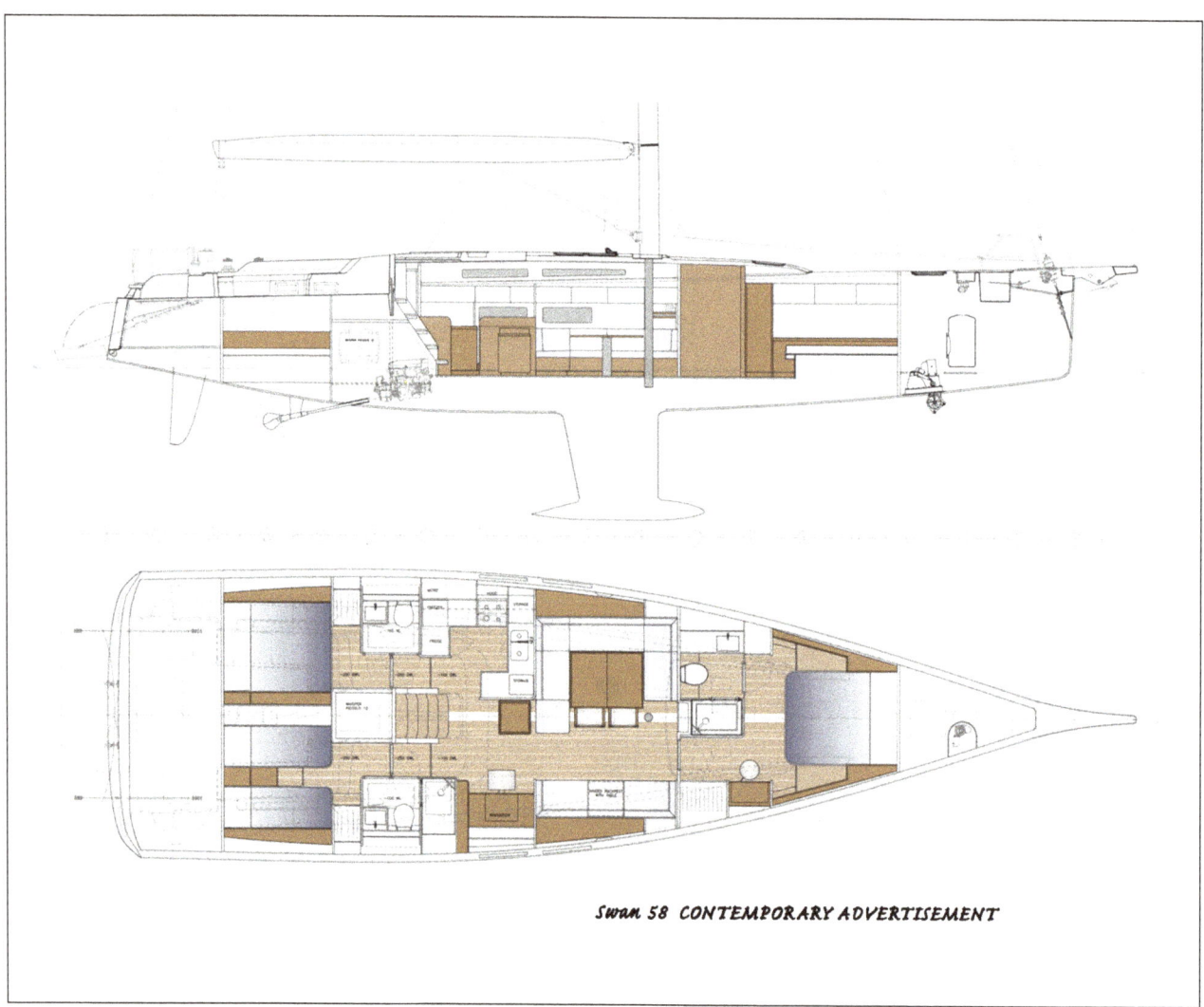

1-18 The Future Is Now. This *Swan 58* advertisement shows an expensive, contemporary fast-cruiser and the computer drawings used to sell her. Swan, a luxury brand, offers this slick, waterized motel with three double rooms, a small office, and a common room/eating area. The *Swan 58* is a far cry from MALABAR X and hardly a boat that would appeal to Slocum.

Although the CCA resisted, racing to Bermuda degenerated into the decadence of purpose-built boats sailed by professionals. I sailed IMPALA on my last Bermuda Race in the millennial year of 2000. We raced against SAYONARA, owned by the owner of Oracle Corporation. She is a maxi purposely designed and built to win such races. For this race her owner had rented the recently victorious crew of the America's Cup, a team nominally based in New Zealand. We had the pleasure of beating them by 16 hours corrected, but it was a silly victory.

CHAPTER 1 ∞ SEA SAILING

I don't know how the Kiwis got home, but I doubt they sailed. *IMPALA* took us home after a sweet passage to San Salvador and six weeks of summer cruising amongst the beautiful and quiet Bahamian isles. But our voyage was an anomaly; big-time ocean racing is no longer about cruising.

During this same period, the second revolution swept the sailing world. Fiberglass reinforced plastic hulls, molded in a single piece, revolutionized construction. Such hulls promised no leaks and, more important, an indefinite life without failing joints. Non-biodegradable, fiberglass reinforced plastic boats needed little care, and less knowledge, for their upkeep. Produced with unskilled labor, they could be inexpensive if produced in quantity. Cheap and easy, they encouraged the sailor to exchange his role as master for the easier role of passenger.

The wondrous invention of fiberglass reinforced plastic changed how sailboats were sold and hence how they were designed. Although the tooling was expensive, costs came down fast if you sold a lot of copies. To sell to the many, designers designed for the average. And because there were not so many experienced sailors, designing for the average meant designing for the inexperienced. The inexperienced were attracted to the fashionable rather than to the seaworthy. The process flooded the sailing world with large numbers of cheap, mediocre boats owned by many who did not know how to use them. Seamanship and its arts languished.

Glass-reinforced plastic construction also drove an emotional wedge between the sailor and his craft, enabling a new breed of yachtsman, who, eschewing personal connection with his craft, saw a boat as a utilitarian object, better to rent than own, better to be dropped off at the dock to be cared for by professionals.

The true sailors, of course, who used their boats the old way, were still there; they were just obscured by the large numbers of the new crowd. They still loved their boats and sailed for family recreation, travel, exercise of friendships, and pursuit of adventure. A few transitioned to fiberglass and a few fiberglass boats were well designed. But most owned wood boats. Their boats remained family traditions, but unlike their family houses, their boats tended to come apart and needed to be replaced every generation. Exquisite as was the craftsmanship that fashioned them from sticks, sawn from trees, into the finest yachts, the metal that held these beauties together caused them, in a couple of decades, to come back apart. And with the fiberglass revolution driving traditional builders out of business, it was becoming hard to obtain replacements or even get the old boats rebuilt.

But it was not quite all over. The same chemistry that created glass-reinforced plastic also created epoxy resins, the near-perfect wood glue. Epoxy bonds wood without pressure or heat. Epoxy-bonding enables the construction of a wooden boat as strong and durable as the trees it came from. Maybe someday biology will allow us to induce trees to grow directly into the form of an actual boat—to be felled and launched directly from the forest. But that is for the future. Today, chemistry has given us epoxy bonding, which allows us to eliminate the great bugaboo of traditional yacht construction—the metal fastenings, whose antagonism to wood causes structural failure. Further, using a technique called cold-molding, epoxy bonding allows the crossbanding of linear wood sticks to make the molded (doubly curved) shell of a boat, strong in two directions, a grail searched for by wood boat builders since that first caveman set out across the water.

With all its promise, epoxy bonding presents the contemporary boatbuilder with a new challenge, that of translating the elegant traditional plank-on-frame boat, a structure that took centuries to perfect, into an appealing epoxy-bonded craft. This is no trivial task, for when glued joints replace mechanically fastened ones, each piece in the boat is changed in how it works and in the shape it must take. All these changes require thought and offer opportunity for art. And the sea suggests that we make our changes without forgetting the hard lessons learned over centuries past.

This book derives from my quest to create a new sailboat construction using cold-molded, epoxy-bonded wood. The balance of Part One relates to my personal experiences sailing boats, my gradual understanding of what the sea requires, and the answers ocean sailing has given me. And then I describe three building episodes spread over 35 years: pioneering epoxy-bonded construction with the *Alerion Class Sloop* and then two other experimental projects, which answered some questions and asked others. Part Three is the result of my third, and virtual, effort. It presents a comprehensive method for building an oceangoing sailboat for a blue water sailor. But before Part Three, Part Two digresses for three chapters into technical discussions of sailboat loads, sailboat structures, and wood as a structural material. Loads, structure, and material set the limits within which we must work. Finally, I give you an afterword about using cold-molding to resuscitate old boats that have become, with age, no longer seaworthy yet remain so desirable that their owners wish to bring them back.

CHAPTER 2

Sympathy with the Sea

SYMPATHY WITH THE SEA brings the sailor joy. Such sympathy comes to the sailor from understanding her moods and accepting her power. Oppose her and, well, she will laugh and you will cry. Sympathize, and your eyes will open to the majesty of her storms, the tranquility of her calms, and the preview of heaven she offers as the sun sprinkles sparkles across her force 3 surface. I say, build your boat to accept her ways and use what she has to offer.

To learn the ways of the sea, listen to what she has to say. You will learn from experience, by sailing over her surface as she expresses her cares and disdains. I was a lucky one. As a child, after summers of Nantucket water, I had winters of Tennessee hills with lots of time, in my father's library, to listen to the sea sing her song through the written voices of sailors past. Their stories were my introduction to later actualities—experiences that brought me to understand a few simple rules for sea-sympathetic boats.

There are many rules; the ones I offer here have to do with oceangoing boats and their construction. The rules are mine,

2-1 **Where sailing begins.** *Rainbows* (known as *Beetle Cat*s elsewhere) in Nantucket, 1954.

not the sea's, and reflect my understanding of the needs of those who sail her. They are derived from my observations at sea. So that you may know their source, to help you judge their validity, I give you a little of my sailing history that explains where they came from.

I learned to sail just as the two radical developments of the mid 20th century hit the yachting scene: the rise of the planing dinghy and the substitution of fiberglass-reinforced plastic for traditional wood construction. By the time I started college, I had learned about sailing dinghies and displacement craft inshore. I knew something about the new plastic construction. It took a good deal longer to learn about the traditional wood construction that I had up till then overlooked. It also took a good bit of time to learn what made an oceangoing sailboat good, because for that, I had to go to sea.

In 1960 my father bought hull #1 of Pearson's first oceangoing sailboat, their *Invicta Class*. She was a 37-foot yawl designed by Bill Tripp to the

CHAPTER 2 — SYMPATHY WITH THE SEA

CCA rule, named *IBO LELE*. This being the late 1950's, there were no standards for fiberglass scantlings.[1] Being their first oceangoing boat, Pearson built her conservatively using scantlings only a bit thinner than if she were wood, ignoring the fact that fiberglass is four times heavier than timber. Her topsides were ½ inch thick and her garboard region thickened to 1½ inches, solid fiberglass. She was certainly strong; she was also highly overweight. Pearson compensated by reducing her ballast and consequently her stability. This upset Tripp, as he had designed her much more lightly built, but since stability was penalized by the CCA rating rule, her lack of it improved her rating. Indeed, *BURGOO, Invicta* #3, won class E of the 1962 Bermuda Race.

I found her sea-kindly despite, and maybe because of, her moderate stability. She handled easily in rough water. As teenagers, my friends and I would sail her out of Nantucket harbor during summer gales. Back then, our east jetty was a submerged one, and 35 knots of northeast wind would kick up a 6-foot sea state that would come right over the rocks. With several reefs rolled in and the little #2 jib set, we would beat out Nantucket's narrow channel, sailing right up to the rocks before tacking. Tripp gave her a sweet helm, and she was responsive to it.

IBO LELE was a good sailing boat, as good, perhaps, as a fiberglass boat could be. She was strong; you need never worry about her hull. She did not leak. Water came in only through purposeful holes—thru-hulls, the stuffing box, and her deck hatches. For her, the bilge pump was an emergency device, not a constant companion. Overall, she was practical, waterproof, strong, and easy to clean. But she was not very sweet inside. She could be cold and damp because of large areas of exposed, uninsulated fiberglass. And, of course, she was overweight. That said, she well fulfilled one requirement for the sailor at sea and taught me my first rule:

RULE 1: A seaworthy boat is strong and watertight.

Having learned *IBO LELE*'s traits, in the summer of 1963, I discovered a very different tradition. My two college roommates and I cruised the sloop *FUN* around southern New England. *FUN* was Ted Hood's first *ROBIN*, built by him four years earlier. She was 40 feet, just a little bigger than *IBO LELE*, and of starkly different build, being traditionally built of wood. She had plenty of ballast and she won races, not from generous handicap, but by being fast and, of course, very well sailed by both Hood and her new owners.

We were three college boys off on a summer cruise, and it was my first time living aboard a wooden boat. Our voyage was true to our vessel's name and without mishap if you will except

---
1. Before *Marine Design Manual for Fiberglass Reinforced Plastics*, written by the engineering firm Gibbs and Cox and published in 1960, there were not only no building scantlings, there was very little engineering data at all. Structurally it was an experimental world. Builders were driven by the promise of a rot-proof, leak-proof hull.

2-2 ***IBO LELE.*** This 37-foot yawl, designed by William Tripp in 1959, was Pearson's first fiberglass oceangoing boat.

our sudden "U" turn just in front of the highway bridge over the Cape Cod Canal. The bridge, it appeared (actual clearance 135 foot), would have taken our mast off at the second spreader. On a second try we made it under with about 70 feet to spare. Such is youth and inexperience.

FUN was a fun boat, though we did pump from time to time; she leaked a bit when driven hard. But as I compared her to *IBO LELE*, I saw something else. On deck and especially down below, her wooden surfaces were far more appealing to look at, to touch, to sit on, and to sleep against than *IBO LELE*'s industrial plastic ones. They were warmer and dryer; they felt and sounded better.

An engineer might argue, "These are just psychological effects—of secondary importance." But he would be wrong. At sea the sailor's boat is his home. His performance, the success of his voyage, and perhaps even his survival depend on his harmonious relationship with his boat. One might argue, "Yes, but most boats never go to sea."

The response is similar: "Even lying to her mooring, the only purpose of a yacht is to give her owner pleasure."

If she gives no pleasure, she has no value. The sailor's pleasure is based on a close, spiritual relationship with his craft. She needs to show beauty and offer tenderness. Down below, only wood will do this.

Sea-kindliness involves more than a comfortable cabin. The motion of the boat is very important. My friend Roger Taylor reminded me of the 1979 Fastnet Race and those sailors who took to their life rafts to escape the battering they were receiving inside their jumpy, jerky boats. Many of the abandoned boats were recovered safe after the storm. Many of the escaping sailors were not.

The drowned sailors did not abandon their floating craft because their cabins were made of cold, wet, unpleasant fiberglass. They did so because the motion of their dinghy style craft was untenable. These unpleasant motions come in four forms. First is the poor steering qualities of unbalanced hulls that are sharp forward and beamy aft, which cause loss of directional control and broaching even in moderate conditions. Second, is pounding, caused by flat bottom hulls with the wind forward of abeam. Third, is the high rolling accelerations generated by the excess stability of hulls with low ballast and beamy, hard bilges.

And fourth, is the high lateral accelerations generated by wave impacts on light displacement hulls, which throw sailors about their cabin, or overboard from the deck their boats. High stability and light displacement are key speed factors. Flat bottoms, combined with sharp entrances and beamy runs, promote planing, which takes speed to new levels. These forms, which increase speed, decrease sea-kindliness. It is no surprise that, as racing has come to dominate offshore sailing, sailboats have become less sea-kindly. Nor is it a surprise that boats lacking sea-kindliness remain tethered in marinas, rather than sailing the sea.

Size also counts. Remember the old adage, "A foot of length for each year of your age." I had the privilege of sailing *QUEEN MAB* to Bermuda in 1982. *QUEEN MAB*, designed and built by Herreshoff in 1910 as *VAGRANT*, for Harold Vanderbilt, is an 83-foot composite steel/wood staysail schooner and an immensely powerful vessel. We left Newport, RI, on a nice, but cold, November day and a day later came to the Gulf Stream. The weather warmed up, the crew cheered up, and some of the bundled clothes came off. I, lying on the bridge deck, look up at the sky and saw a perfect halo around the sun. "Enjoy the weather," I said [while you have it]. Twenty-four hours later we were in a northeaster under fore-staysail. We had a man at the helm, but it was unnecessary, she was basically steering herself and broad reaching before the 15-foot seas with a bit of heel and little roll. We were going fast and as *QUEEN MAB* hit hull-speed she began to bring the larger waves aboard over the stern. I suggested handing the staysail, but the Captain said no. A little later it blew out and the ship settled down. We were in the Gulf Stream and the wind was force 8/9 and as I stood on deck, holding onto the fore-staysail boom, I was reminded of being on Nantucket's Steamship wharf during a summer hurricane, so smooth and powerful was the *QUEEN*.

Quite in contrast, a few months later, I was steering *BULLFROG* in California's Santa Barbara Channel. *BULLFROG* was a state-of-the-art ocean racer from San Francisco. We were running in force 5 on a dark night, a four-foot sea was running. We were sailing by some of the oil rigs there and the glare of their lights hampered visibility of the sea in front of us. Suddenly an anchor buoy appeared directly ahead, and I made a quick turn of the helm to dodge it. Immediately we found ourselves heeled over 45° and without steering control as our rudder was in air. I was taken by surprise, so the skipper pops up and releases the main sheet. The main boom was in the water, but was short enough that the sail was eased, depowered, and we righted, gaining steering control as we did so. All quite startling in 20 knots of wind.

Easy motion (and steering) is essential to security at sea aboard a small sailing vessel.

RULE 2: A sea-kindly boat has a wooden cabin, and an easy motion in a seaway.

The fiberglass marketer was not concerned with sea-kindly motion one way or another, but his response to the need for the ocean/human interface to be wooden was to build a fiberglass boat and then build a wood boat inside of it. On the exterior the sea was exposed to the fiberglass. In the interior, the crew was exposed to wood. In the late 1950's such boats began to be built, perhaps the most famous being Hinckley's *Bermuda 40*, again designed by Bill Tripp. This boat, a fiberglass boat that looked, and to some extent felt, like a wood boat, quickly established

CHAPTER 2 — SYMPATHY WITH THE SEA

2-3 *WIND*. Sister to Carter's first Fastnet racer, mentioned in chapter 1, *WIND* was a *Medalist*. Thirty-two feet long and beamy, she was a good sea boat with plenty of room for a young couple. She even had a forward cabin with a door that could be shut when one was angry. She had a nice wood interior concealing her fiberglass hull, which was very satisfactory until it became necessary to know what was being concealed behind the ceiling! Here she is sailing out of Marseille past Ile d'If of *Count of Monte Cristo* fame, December 1964.

Hinckley as America's finest boatbuilder in the fiberglass age.

In the fall of 1964, I flew to Jutphaas with a new bride to take delivery of my first ocean-going sailboat, a LeCompte *Medalist* (sister of the *Medalist* discussed in chapter 1). She was another Tripp design, a 32-foot fiberglass sloop, a smaller version of the *Bermuda 40*. She had a finely fitted interior of teak and mahogany concealing her fiberglass hull. To her crew she was almost a wood boat; to the sea she was strong and watertight.

We were headed for the Mediterranean. Our motive was a desire to see the world, by sail. Perhaps I was subconsciously inspired when, in that library of my father's, I read Goran Schildt's book about he and his wife sailing their ketch *DAPHNE* across France to the Mediterranean. The Schildts were middle-aged; the French considered us children. At 21 and 19, we thought of ourselves as explorers, out on our own, in our lovely *WIND*, away from home and family for the first time. After a bucolic fall cruise through the inland canals, we arrived at Marseille in December. It had become cold. Our unaware assumption was that the Mediterranean was tropical. In truth, the winter there is wet, cool, and stormy.

Typical was our sail out of Monaco on December 10, 1964. We had awaited a good day, and there was a light wind. I set full sail, main and #1 genoa, and slipped out of the harbor, too proud to use the engine. The breeze began picking up, so I shortened down to the #2 genoa. Then I took a reef, then reduced to #3 genoa, then another reef, then reduced to working jib, then #2 jib, another reef (*Wind* had roller reefing), and finally about 3 hours later we found ourselves slopping along in force 8 under trysail and storm jib. The sun went down at 4 into a cold sea. The next morning the wind began to moderate. I went through all the same changes in reverse before coming into Elba more or less becalmed under genoa and full main. This is how ignorance develops experience. I got very good at changing headsails that winter and experienced with heavy weather. We began all seven of our passages in calm, but five ended in gale. It was a thorough schooling.

Like all cruising boats, *WIND* was more than just a sailing vessel. For almost a year her 500 cubic foot interior (a third the size of a "tiny house") was our home. Living aboard puts one in intimate contact with the boat and her surfaces. *WIND*'s wood finishes and ceilings made that contact pleasant. Her cozy interior kept our spirits high through our heavy-weather classes. She was a source of the pride we took in our voyage. Such pride is not vanity, but rather an important component of seamanship, for a crew that is proud of their vessel will sail her better. We began discussing our Atlantic passage home to Nantucket.

Six weeks in Bizerte awaiting spring was as much as we had patience for, so come mid-March we headed for the Palma de Majorca, 380 miles closer to Nantucket. As with our earlier passages, we found ourselves beating to windward against force 7 to 8 with storm trysail and storm jib—a little under-canvased. On the third day out a squall, force 10, solved the undercanvassing—the boat took off at a mad dash. We burst over a steep wave and came down hard on the other side with a crash. Both thrilled and frightened, it took me a moment to notice that the leeward aft lower shroud was loose; its chainplate had pulled up through the deck about 2 inches.

CHAPTER 2 SYMPATHY WITH THE SEA

The masts on CCA style boats back then were built for the sea, meaning they were over-built, so our rig was not in danger of coming down, at least not yet. But what went wrong? What was the extent of the failure? What might happen next? Because *WIND*'s beautiful wood interior covered her industrial plastic structure, I could not see to assess the damage. I had neither the tools nor the skills to open her up at sea and uncover what had gone wrong. Ignorance begets anxiety, anxiety generates stress, stress causes fatigue. A fatigued crew makes mistakes. Mistakes will turn a seaworthy vessel unseaworthy. Left to their own devices, Poseidon and Aeolus will produce doubt enough for the sailor; his boat should reassure him, not hold secrets from him. We altered *WIND*'s course for the nearest port and limped into Mahon, Menorca, through two more days of gale.

Our problem did not seriously endanger the boat, though it could have. A failed upper-shroud would have meant the loss of the rig. The bigger problem was that the failure frightened us. The structure being hidden, we could not fix it; more importantly, we could not understand it. We ended our voyage in Mahon with a call to LeCompte and the sale of the boat.

RULE 3: Spaces and structures without access are unseaworthy.

I returned from Europe to start architecture school at MIT. Four years at MIT gave me a strong foundation in material science, structures, and the art of design, and for what it might be worth, a degree in architecture. Meanwhile, as a pastime, I had built a foam-core fiberglass boat and, separately, put a wooden interior into a fiberglass pocket cruiser.

As mentioned, during the 1960's, a battle between fiberglass and wood construction raged in the hearts and minds of the yachting public. By 1975 it was all over. Fiberglass had won the sailboat wars. The boat-buying public had expanded five-fold and most of it had decided that fiberglass was the only sensible way to go yachting. Simultaneously, the ocean racing world abandoned its traditional sea-kindly model based on traditional craft for a new model based on the planing dinghy. When the IOR rating rule was instituted to promote the change, the wooden, CCA, inspired ocean racers developed from the 1920's through the 1950's became obsolete.

As their prices dropped dramatically, they started showing up in Nantucket, no longer "gold platers" but rather cheap boats affordable to young sailors. Some of these young sailors moved their boats between New England, summers, and the West Indies, winters. To passage back and forth they needed navigators. My ability with celestial navigation got me a berth on several.

Two of my delivery trips were aboard my brother's *ASPARA*. She was an unusually pretty Alden ketch, 56 feet long, built in 1927, mahogany planking over oak frames. She was sea-kindly and fast, as one might expect of a John Alden oceanracer.

On our first sail, a purchase-delivery home to Nantucket from Florida, we left Ft. Lauderdale at 2 PM in light air with me as navigator and Michael Walker as skipper. Squalls appeared that afternoon. After an hour of downpour and gusty wind, the front went through leaving a brisk southwesterly behind it. *ASPARA* had an off-center prop that would start her singing at 7½ knots. At 9 knots she sang loudly. We carried on, full in the Gulf Stream, and by our sun fix the next afternoon we were off Charleston, 275 miles on our way!

ASPARA was a lovely boat, but she was iron fastened. She dramatically demonstrated the primary weakness of traditional plank-on-frame construction: metal fastenings. All metal fastenings are antagonistic to wood. Iron ones are particularly so, especially when used with the usual oak frames. The acid in oak rusts the iron. The iron's rust encourages the oak to rot. At 45 years of age *ASPARA* was weak and, in truth, no longer seaworthy.

The bronze screws Herreshoff popularized are far less antagonistic than the iron nails used by Alden's fish-boat building yards; copper rivets fancied by the best British yards are even better. But metal and wood are fundamentally antagonistic. Metal, being much harder than wood, causes crushing of the wood fiber. When the structure flexes, any sharp metal edge—think the threads of a screw—becomes a saw tooth, with a similar action to that of a contemporary Multitool. The lifetime of a metal-fastened boat is 20 years if iron fastened oak, to 45 years if Herreshoff (structurally ceiled) or English (copper riveted) built.

Metal conducts electricity, so changing electrical fields (think alternators) induce electric currents that chemically decompose the wood. Electrolyzed wood has the strength of mink fur. Metal also conducts cold well, drawing atmospheric moisture deep into the interior of the wood and condensing it, promoting rot. Altogether an unhappy situation. As my friend Ed Cutts told me at his pretty yard in Oxford, Maryland, "The old boats, the really old ones, don't use metal."

RULE 4: Metal fasteners destroy the wood boat they hold together.

I learned another weakness of traditional plank-on-frame construction delivering boats to the West Indies. One was the Rhodes sloop *PIERA*. She was built by Abeking and Rasmussen to very high standards in 1955. Although better built than *ASPARA*, she leaked because her hull was not strong enough to carry the loads imposed by her high Marconi rig. Although the stresses on a sailboat hull mostly run fore and aft, loading due to the rig and ballast puts high athwartship loads on the hull especially adjacent to the mast. With plank-on-frame construction these loads are imposed on the frames, in addition to the loads from crossbanding the planks. In *PIERA*, as in many highly ballasted, tall-rigged boats, the frames were not up to the job.

CHAPTER 2 — SYMPATHY WITH THE SEA

We left Nantucket, one late November afternoon, headed for Bermuda and beyond. We got around Great Point and were settling the ship down for the passage when the skipper called me to look at the hull, which was exposed to view in the head compartment alongside the mast. We had a gentle breeze, force 3, and we were close reaching into a regular swell. Each time the boat settled onto a swell, one could see thin streams of water squirting through the seams of the planks. None of us were experienced enough to accept this as normal, so we turned around and sailed back across Nantucket Sound to *PIERA*'s repair yard at Mattapoisett. At Cross Rip a gentle snow began to fall and by the time we arrived at Mattapoisett, well after dark, the northeaster was beginning to blow.

Brownell trailered us out early the next morning, put in some additional caulking, reminded us of the primacy of the bilge pump, and sent us off again. We made it to Bermuda—and farther on.

But *PIERA* still leaked. And when pressed, she leaked a lot. There was no getting around her weakness, which was an inherent fact of her construction. *PIERA*'s leaking was caused by the failure of the connection between her frames and the floor timbers that supported her mast step. This connection is traditionally made by bolting the floors to the frames. The bolt holes substantially weaken the frames, which are undersized for the load in the first place. Frames are sized to crossband the planking and resist its athwartship swelling; their section is more or less constant throughout the boat. They need to be increased 2 or 3 times to carry the mast loads. Usually they are not, often because there is not enough room, under the sole and on top of the keel, to do so. The frames break at the floor bolt holes. Then, when the mast step pushes down and the shrouds pull up, the planks separate because a broken frame won't hold them together. The sea squirts in.

The gentle vision, on that snowy afternoon, of these quiet streams coming through *PIERA*'s planks brought home to me the necessity for what I have come to call "hoop" strength, or tensile capacity, transversely all around the hull. A seagoing boat needs it, especially at the mast.

Rule 5: A seaworthy boat has "hoop" strength right around her section.

In 1976, I sailed the S&S-designed yawl *OTAIS* across the Atlantic. She was a 42 foot version of *FINISTERRE*, a heavy, beamy centerboarder, very well built by Walsted in Denmark and not too old. She was a lovely boat and without speedometer, depth sounder, running water, hot water, refrigeration, or radio telephone.

Once departed from Nantucket, we sailed her dark because her batteries would not hold charge. We navigated with taffrail log, compass, and sextant, using a dry cell–powered transistor radio to get the time signal. After 14 days of strong southwesterly September winds, 1,800 miles on, we came to Flores, the first of

2-4 *OTAIS*. The design of *OTAIS* (now *AQUILA*), a powerful boat with simple gear, is based on Olin Stephens' *FINESTERRE*. *OTAIS* kept us happy at sea. Here, we are about 1,000 miles from each of Nantucket, Flores, and Cape Farwel, Greenland.

the Azores, where we were greeted by the local whaleboat and escorted into Santa Cruz. They tied us up near the foot of their boat ramp with lines spider-webbing us to a circle of semi-submerged rocks, protected in the prevailing west winds by the bulk of the island. As we came ashore, the port captain instructed me, "If the wind goes east, leave. Don't say goodbye. Just leave!"

Of course, we had no power hookup, water was by jerry-jug, and communication was by Portuguese telephone operator. The island was connected to the other Azorean islands by plane three times a week and by ship once every two weeks. At Flores, dockage was managed by winching boats up the stone ramp above the reach of easterly seas. She had no boatyard, at least by New England standards.

CHAPTER 2 — SYMPATHY WITH THE SEA

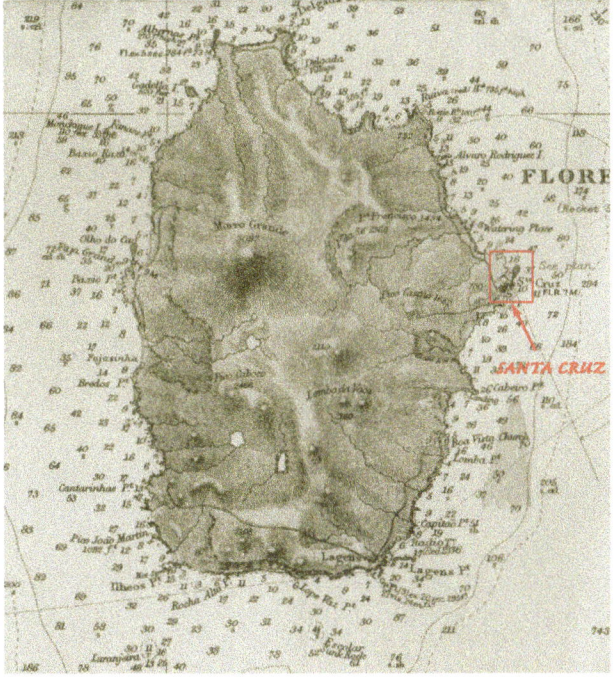

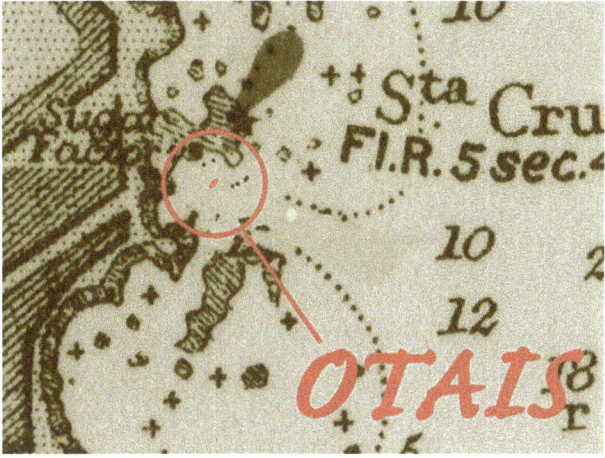

2-5 Harbor of Santa Cruz! *OTAIS* is webbed into the "harbor" at Santa Cruz, Flores. Only a few semi-submerged rocks separate the cove from the Atlantic. The wind is in the north, so the cove is tenable for the moment.

No problem. Because *OTAIS* was robust and simple, we needed no service. We were free to meet with the world's last whalers working from their rowing boats and enjoy the solid society of islanders who had lived together semi-isolated for four centuries. They were fine people. For us, it was a good time.

A day later we did leave, with proper goodbyes, and continued our voyage to Palma de Mallorca. I had learned that to visit the world's out-of-the-way places, one needs a boat like *OTAIS*. A strong, pretty, cozy boat that can take care of herself (with your help.) She will bring you across an ocean, to faraway islands elsewise inaccessible. An ocean cruiser does not want *fancy* equipment. She wants equipment that *works* and keeps working and that, if broken, is fixable with the tools she carries on board. Robustness is maybe the most important of all qualities of an oceangoing sailboat. On the ocean you are a long way from the boatyard, and once started, you do not want to turn back.

RULE 6: A seaworthy boat has simple, robust equipment repairable at sea.

Here are the rules in summary:

RULE 1: A seaworthy boat is strong and watertight.

RULE 2: A sea-kindly boat has a wooden cabin, and an easy motion in a seaway.

RULE 3: Spaces and structures without access are unseaworthy.

RULE 4: Metal fasteners destroy the wood boat they hold together.

RULE 5: A seaworthy boat has "hoop" strength right around her section.

RULE 6: A seaworthy boat has simple, robust equipment repairable at sea.

Minding these rules won't guarantee a happy oceangoing boat, but you definitely won't have a happy boat if they are broken.

CHAPTER 3

Pioneering Cold-Molding

WITHOUT ANY PARTICULAR plan or intention, much of my life has been spent thinking about how to build boats. I began sailing in wood boats, but I came to wood boat building late. Although my first memory of the workshop is of watching my father apply resorcinol glue to the chines of a plywood pram, I was early introduced, in that shop, to the radical new material—fiberglass-reinforced polyester plastic. Back then, laminating resin did not come pre-mixed in quart cans. Rather, we mixed our own concoctions from a drum of polyester with portions of permanganate, monomer styrene, and chemicals whose names I can no longer remember, measured out on a balance scale. We set it all off with methyl-ethyl-ketone and mushed it into and onto glass fabrics stretched over a form and came up with a hull that would not rot!

While at MIT, in the mid 1960's, I built the proverbial boat-in-the-backyard. She was a foam-core, fiberglass 26 foot version of Ian Procter's *Tempest*, complete with a lifting fin keel with bulb ballast. I got her finished, launched, and sailing. She was a structurally sound and totally awful boat, occasionally fast.

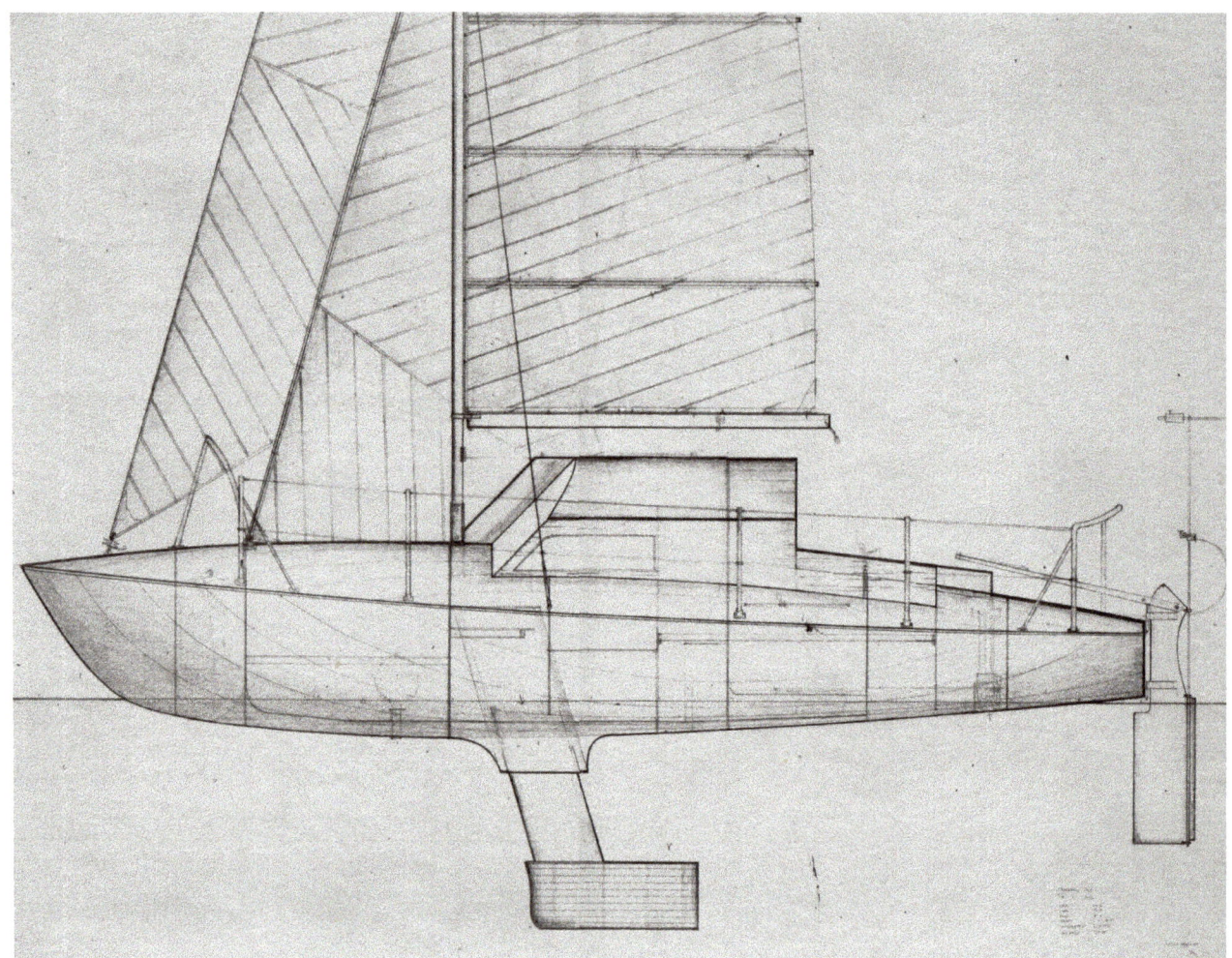

3-1 *JET ROCKET*. Built in my backyard while at architecture school, *JET ROCKET* was based on Ian Proctor's *Tempest*. Blown up 20%, she had all the avant-garde elements of the day, extreme planing hull, lift keel with bulb, outboard rudder. Showing my confusion between going fast and going places she had a strong whaleback deck and an awkward house to give her room below. She was foam/fiberglass construction and weighed about 2,000 pounds, of which half was lead.

As I grew older, my sailing experiences brought me to better understand and appreciate wooden boats. I had learned the advantages of traditional wood construction—strength, light weight, and domesticity. And I had learned the source of its weakness—metal fastenings and the difficulty of joining the linear material to form a two-dimensional one, molded into a doubly curved surface. I saw that fiberglass overcame these weaknesses but that it was compromised by its low strength, poor thermal characteristics, and inherent unpleasantness. The time was ripe for a new technique for building boats. The idea of cold-molding began to stir.

It began to stir in my mind, it began to stir in my brother Edward's mind. Between us we got it to stir in my father's mind. The story of how the three of us stirred it up in the public's mind is the story of the creation of the *Alerion Class Sloop*. The story goes like this.

The story of the Alerion Class Sloop

Take yourself back four decades to 1977. Place yourself in Nantucket, a pretty little spot, a sleepy summer resort grown up amongst the abandoned structures of the great 18th century whaling port. By 1977 the times have changed. Wealth has given way to poverty, grandeur has given way to quaintness. The 20th century has adapted the quaint for the summer pleasure of folk from new cities. And even that is changing. The beautifully constructed wooden boats that have sailed Nantucket's waters for the past century are mostly gone. The Mower-designed Vineyard Sound Interclubs, the Starling Burgess–designed Yankee One Design fleets are no more. The hot-molded plywood *Flying Dutchman*, 32 of which scooted around Nantucket harbor in the late 1950's, has disintegrated. Carl Beetle's 12-foot *Beetle Cats*, known locally as *Rainbows* because of their colored sails, are a dwindling fleet. Now, in 1977, Beetle's shop in Padanarem, which built whaleboats for the Nantucket whaling fleet, is probably the last shop in America producing a stock wooden boat. Now, the sailboat wars are over; fiberglass has won. The sound of the plane and the caulker's mallet has been replaced by the screech of the fiberglass polisher.

The excitement of the moment is that my father, Teeny Sanford, a top round-the-buoys man in his youth and later a blue water cruiser, has, with his friend Pilly Mills, just bought a new sailboat in Florida and brought her to Nantucket. Called ALERION, she is the prototype of the *Alerion 26* class, newly created by Halsey Herreshoff. She is a good-looking, 25 foot keel sloop well built in fiberglass. Teeny and Pilly will use her for pleasure sailing Nantucket's harbor, even though her 43 inch draft is a little strong for the shallow waters. She was a pretty boat. Her looks so intrigued my father that he wanted to know more about her background. A few years before, I had seen Nat Herreshoff's (Halsey's grandfather) ALERION, on display at Mystic Seaport, and studied her design and history. I tell him the story:

> In 1912 Nat Herreshoff, Halsey's grandfather, designed and built ALERION (his third boat of that name and sometimes referred to as ALERION III, although that was not on her transom) for sailing

in Bermuda, where he wintered to avoid the New England cold. She was the 24th of 29 boats Captain Nat personally owned. Herreshoff was 64 at the time. He sailed *ALERION* in Bermuda for 8 winters, then brought her back to Bristol for summer use. Indeed, he sailed her until he could not. As he reports, "In the following fall [1929] I was not reliable on my feet and decided to give up using boats entirely."

He was then 81. At the end of his life, Herreshoff said of his 26 foot sloop, *ALERION,* "She was a very satisfactory boat"—high praise from the world's greatest yacht designer.

What made *ALERION* such a "very satisfactory boat"? In 1912, both Herreshoff and the Herreshoff Manufacturing Company were at the peak of their powers. Nat Herreshoff had been designing sailboats for nearly 50 years and Herreshoff Manufacturing Company had been building them. He modeled *ALERION* as the great age of sail was coming to a close. By then, the shape requirements of a fast, seaworthy, comfortable sailboat were known. Back then, people used sailboats intensively, both for work and for pleasure. They knew good boats and were selective about them. Herreshoff was amongst the most selective.

He knew there are two ways to make a boat go fast. She could be narrow like a rowing shell and slice through the water, or she could be flat bottomed like a scow and skim over the top of it. He also knew neither shape produced a sweet sailing boat, the narrow boat because it lacked ability to carry sail, the flat bottom boat because it pounded in rough water.

For his personal sailing boat, Herreshoff eschewed both extremes; instead, he began with Edward Burgess' 'compromise sloop' shape, whose model had moderate beam, short overhangs, and a wine glass midship section. This model incorporated the subtleties developed by fishermen, pilots, and yacht designers during the 19th century that lower the resistance of a sailing craft. These included the aft sloping midship section, the sharp entrance, optimum prismatic coefficient, and broad, smooth run.

Seeking to gain the advantage of light displacement (speed) while avoiding its disadvantage (pounding of a flat bottom in a seaway), Herreshoff modeled *ALERION*'s turn-of-bilge to begin just at the waterline. For practical purposes the underwater portion of her midship section is triangular, giving minimum cross-sectional area (displacement) for the angle of dead rise required to avoid pounding.

CHAPTER 3 PIONEERING COLD-MOLDING

Herreshoff modeled the lightest (fastest) hull still possessing good sailing characteristics in rough water.

This innovation of Herreshoff's is not well recognized, nor has it been much imitated. The later International, CCA, and RORC rule boats have turn-of-bilge well below the waterline. They are heavier and slower but no more seaworthy than the *ALERION* style designs of Herreshoff, which continue to vex competitors on the race course today.

My father and Pilly were experienced, well-informed yachtsmen of the old school; this tale need not have been told them but for the fact that Herreshoff's *ALERION*, built before they were born, had fallen into obscurity. I continued:

> Herreshoff sold her in 1929. She had a succession of owners. In 1970, your friend, Ike Merriman, inherited her from his father in poor condition. To preserve her, he donated her to Mystic Seaport in 1964. Missing her, in the mid '70s, he asked Halsey to create a new one. The result is your new boat!

There is much discussion around the Sanford dinner table after *ALERION*'s arrival in Nantucket about good sailboats, wooden sailboats, the problems of wood construction, and the then new cold-molded construction technique. My brother Edward, who had bought *ASPARA*, the 1927 Alden Ketch, in 1971 and sailed her in the first Opera House Cup Race in 1973, brings to our father's attention the advantages of cold-molded construction.

> Cold-molded wood construction actually is quite a bit stronger than fiberglass, extra strength that can be utilized for extra ballast and better performance. Cold-molded construction does not depend on metal fastenings, the destroyer of the traditional plank-on-frame boat. Cold-molded boats are in fact tougher than fiberglass boats and easy to maintain. Maybe most important, but difficult to measure, in a cold-molded boat, the glory of wood surrounds the sailor. People have lived in close proximity to wood for thousands of years. They like it; it makes them feel good. The finest fiberglass boats try to get the wood feel by building a wood boat inside their fiberglass hulls, but this construction is not only heavy, it does not convince. A wood liner is different from the interior of an actual wood structure.

My father, Teeny, has been a pioneer in fiberglass boat construction and he knows its limitations. Connecting the dots, he proposes, "This new boat of mine is a nice fiberglass boat. Now, let's build a real *ALERION* with cold-molded construction."

So the project was born.

Building the *Alerion Class Sloop*

We started Sanford Boat Company, Inc., with two goals: to replicate Captain Nat's *ALERION* and to build her using the new process of epoxy-bonded wood including a cold-molded hull. Our idea was to stay as close as possible to Herreshoff's design but to use the new technique to reduce her upkeep and increase her longevity. We called our boat the *Alerion Class Sloop*, acknowledging Herreshoff's design and her new life as a class boat.

Both the building of replicas and cold-molded construction were on the wild side in 1977. First, in 1977, most builders were offering their customers cheap racing sailboats typified by the dinghy-hulled *J-24* based on the *Flying Fifteen* concept. Rather than follow them, we told our customers, "Now it's your turn" [to sail a boat designed and built by the master for experiencing the pure joy of sailing]. We were offering a sailing elegance forgotten in the rush of the modern world.

Second, by 1977 most builders were exploiting fiberglass construction and convincing their customers that fiberglass construction made a strong boat that was "maintenance free." We were countering these ideas by pointing out the strength of cold-molded wood (**RULE 1**), which allowed a boat more ballast. And we explained the durability of cold-molding, which meant that an *Alerion Class Sloop* could be depreciated over a lifetime, offsetting its higher initial cost. We emphasized the importance of beauty and elegant design (**RULE 2**) by pointing out that the purpose of a yacht is to bring its owner pleasure. No low price will mitigate the failure of an uncomfortable or ugly yacht.

Needless to say, our gambit ruffled a few feathers particularly amongst the old-line Herreshoff aficionados—we lacked membership in that club. To the credit of their position, our move was audacious. Cold-molding was a new technique; the books had yet to be written, and actual builds were few. We were new, and we found ourselves on the frontier of a new world in which we were, by necessity, our own guides. In the course of making wonderful discoveries, we also made mistakes. My father had taught me long before: it takes three tries to make something new. On the first try you are searching, you know not quite what for. On second effort you make a pretty nice job of your new idea. And on the third you perfect it.

And it did take us three tries. With our first Alerion Class Sloop we copied traditional construction, mostly substituting glue for wood screws and bolts. After its completion,

3-2 Two *ALERIONS*.

Top: **N. G. Herreshoff's *ALERION*, sometimes referred to as *ALERION III* because she was his third boat of that name.** Mystic Seaport claims the photo is 1939, but it looks like Captain Nat himself sailing her, so it might be before 1929. She sails with a cut-down jib set above her working sail.

Bottom: **Alerion Class Sloop #21, *KITHERA*, in 1982.** She has just been launched and test-sailed, awaiting delivery to Seal Harbor, Maine. Today, 38 years later, she remains in her original owner's family. She was the first of the *Alerion Class Sloops* with teak decking and as a special order also had teak outer planking.

Maynard Bray paid us a visit. At that point in time, Maynard was a master boat builder whose experience was of particular interest because he had accomplished the restoration of Capt. Nat's *ALERION* for Mystic Seaport.

He examined our immature first effort and asked me how I was going to do the hull/deck connection on our next hull. I did not have a good answer. I realized he did not have a good answer either. Maynard is an honest, direct and constructive teacher. I realized that if he had known the answer, he would have offered it; and if he didn't have the answer it was because no one had the answer. He had only the question. And the exhortation, "You had better figure it out."

It slowly dawned on me that there was a whole new world out there, waiting to be discovered. A world of opportunities to use cold-molding and epoxy bonding to advance traditional boatbuilding. Each joint, each element, was open to rethinking that would reward the builder with a stronger, more elegant and easier to build boat. It began my search that culminates four decades later with this book.

We proceeded to invent, beginning with hull #2 and coming to fruition with #3, the first proper *Alerion Class Sloop*. Although most of her parts had analogues to those of Herreshoff's *ALERION*, very few pieces of the *Alerion Class Sloop* were actually the same as *ALERION* parts, subtly differing in shape and fit and radically differing in action. While, to the layman, our boats looked the same as Herreshoff's original, in actuality they were quite different. Different in many little ways and very different in three big ways.

Firstly, our hulls were planked of four layers of thin wood, starting with a ¼ inch fore-and-aft, spiled inner layer, then two ⅛ inch layers laid diagonally, finished with a fore-and-aft ¼ inch layer on the exterior. Planked this way, two-thirds of the hull material ran fore and aft aligned with the principal forces on the hull. And our hulls could be finished bright, showing off the beautiful shape and pattern of the carvel planking.

For the non-boatbuilder, first a note on planking. Planks look straight on a finished boat, but when flattened out, they take their so-called "true shape," which is banana or "S" shape. There is a direct relationship between the true shape of the planks and the three-dimensional form of the hull, but that shape is not intuitively obvious (at least not to me). Finding the true shape of the plank is called "spiling." The spiling can be documented by a set of offsets. Once the spiling is determined, it is used to cut correctly shaped planks out of boards. Traditionally, determining true shapes is done one plank at a time, as the planking gang planks up the boat. It is time consuming.[1] Planks of normal

---

1. Some shops use double-thick plank stock, spile it, and then re-saw from two identical planks, one for port, one for starboard, so spiling two planks for the effort of one.

3-3 Spiling.

The true shapes of the carvel planking for the *Alerion Class Sloops*. The inner and outer layers of planking used the same shapes. We used 16-foot stock, so the planks required one or two butts. The butts on the inner and outer layers were staggered. For the inner layer we used eastern cedar and, on one boat, cypress; the outer was Honduras mahogany, sapele, or teak.

There are 13 planks in all. The sheer plank had the swelled section that provided a rub rail. It was tapered but not spiled; rather, it was edge set and twisted into place, a process that involved quite a bit of force and many clamps.

All the planks were tapered along their length with a common taper function. This meant the top and bottom curves were not parallel. Planks 9 through 13 were 50% wider than 1 through 8 as there was less shape in the bottom region than in the topsides. The saw table was 32 feet long and had a flexible batten to guide the Skill saw. Four layers of planking were prepared with butt joints in proper place and stacked on the table. Then the batten was fixed to the correct curve using a table of offsets for the spiling. The top curve was cut and then the batten was moved as required by the taper function to cut the bottom edge. Then we moved on to the next plank. Using this rig, two men could cut a gang of planking in a day. We'd cut three boats' worth and then put the table away for three months.

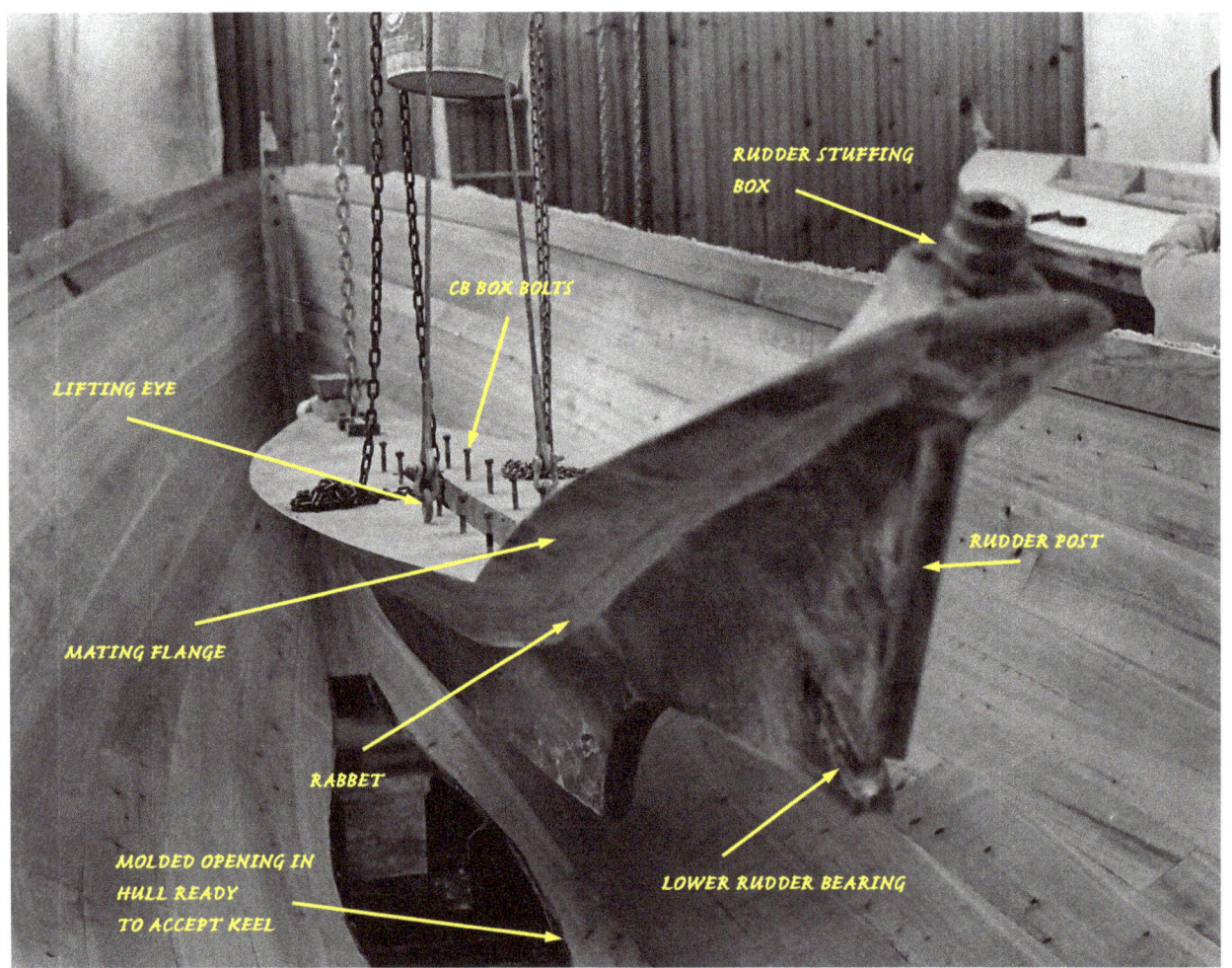

3-4 The ballast used as structure. The *Alerion Class Sloop*'s keel was molded of epoxy fiberglass and lead ingots. First, a 3/16 inch shell was molded of epoxy mat and woven roving. Into that, a mixture of epoxy, chopped fiber, and litharge was poured as a mortar to hold shaped lead ingots in place. Finally, a smooth top surface was cast, which was to be the surface of the bilge. The rudder bearing and stuffing box were cast in place, and after the keel was pulled from the mold, the centerboard box bolts, and lifting eyes were installed.

thickness can compensate for imperfect spiling, by edge-setting, or bending them edgewise, and forcing them into place. This will not work for the thin planking used in the cold-molding process, because the thin stock will buckle when edge-set. The planks must be cut precisely, especially if the hull is to be varnished bright. The precision, of course, is expensive.

With the *Alerion Class Sloop* we spent our time getting the shapes using the traditional method, in situ, with a spiling batten (called a batten, but in actuality a straightedge). Once we

CHAPTER 3   PIONEERING COLD-MOLDING   51

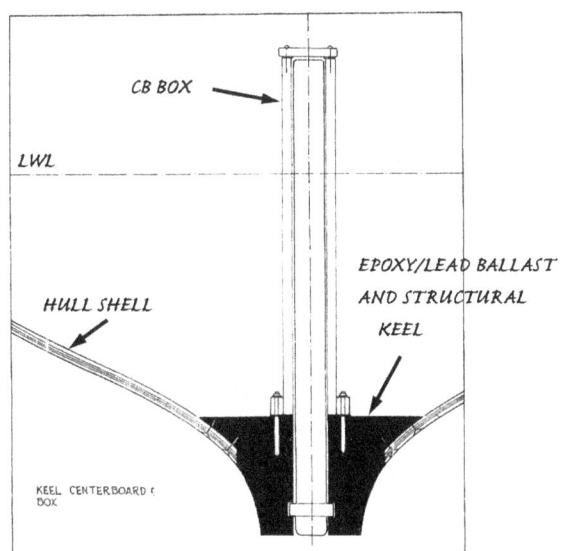

3-5 **Keel/hull joint.** The *Alerion Class Sloop*'s keel drops in like a keystone in an arch.

had the spiling, we made our economy by using it over and over on our run of identical boats. Since our interior layer and outer layer used the same spiling, we placed our plank stock in a stack of four layers—inner port, inner starboard, exterior port, exterior starboard—and got four planks from two long cuts, one for the top edge and one for the bottom.

Secondly, after the hull shell was complete, there was the question of how to build the keel. We had built the first boat more or less conventionally, hanging her ballast keel from floors with keel bolts. We only needed to do it once to ask the question, why, after eliminating the framing and distributing the sailing loads throughout the shell, did we want to re-concentrate those loads into floors and hang the ballast from them with a few large, highly stressed bolts?

Inspired by L. Francis Herreshoff,[2] we realized we could use the ballast itself as a structural member, replacing six wooden elements of Herreshoff's *ALERION*—keel, deadwood, floors, mast step, centerboard box foundation, and rudder post—with a single lead casting cased in epoxy fiberglass. To my knowledge this had never been done before, but it worked as planned. It had the happy result of eliminating the large unstable timbers of keel and deadwood, many crevices that trap water and dirt, and many bolted connections. It produced a clean, plastic-surfaced bilge, the one place on a boat where you want plastic. Wood is at its least advantage in the bilge, where its beauty is unseen, its light weight is irrelevant, and its propensity to rot due to wetting and drying is a vulnerability.

To attach the keel to the hull shell, it was formed with a rabbet and a wide flange into which the wood hull shell would fit. The keel was dropped in from above, somewhat like a keystone into an arch. Since a cold-molded shell has continuous transverse strength derived from its diagonal layers, the keel could be simply glued along the entire surface of the rabbet. The keel was a collection of lead ingots set in a mortar of epoxy reinforced with fiberglass. Its epoxy surface was easily glued to the hull shell.

---

2. Herreshoff, *The Common Sense of Yacht Design*, pp. 73–74.

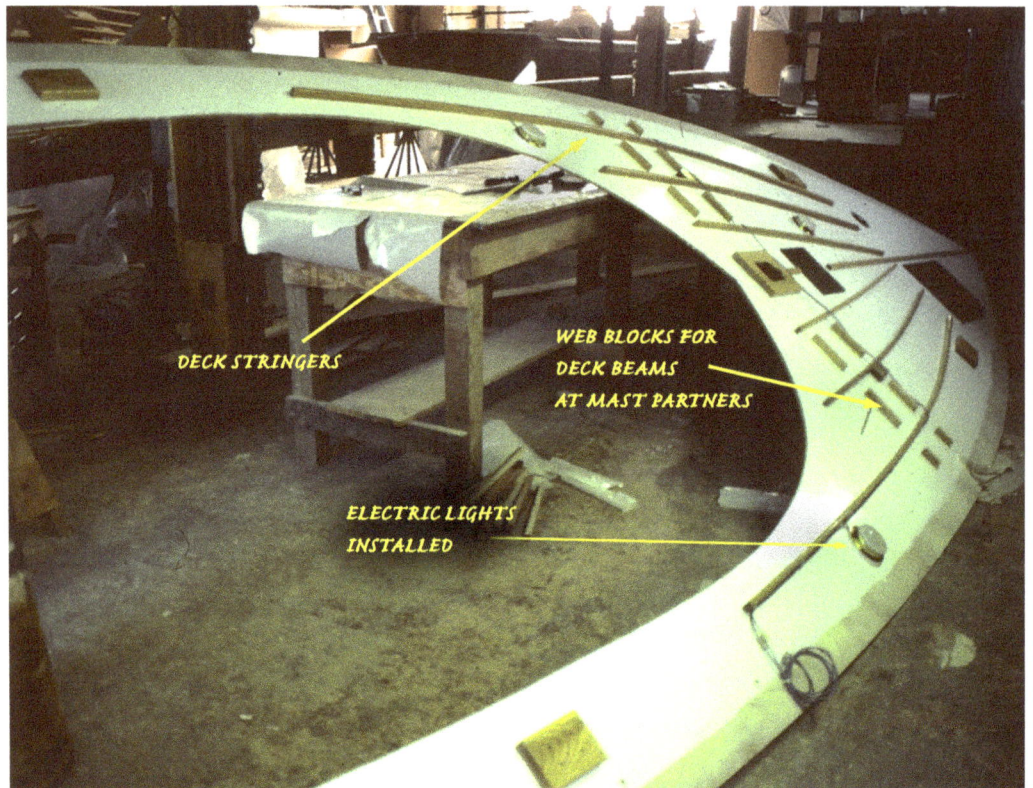

3-6 Deck framing. The *Alerion Class Sloop*'s deck is framed with light longitudinal stringers. After it is in place on the boat, deck beams are bent into place at the partners and cockpit openings, forming, with the stringers and spacers, a vierendeel truss. The deck beam/truss is visible at the aft end of the cockpit.

The keel load was distributed not through highly stressed metal bolts tying the ballast to floors, that were then bolted to frames, that were then screwed into the planking, but rather across 13 square feet of glue joint. Each square inch of joint carried about two pounds of load, and only then if the boat is aground or upside down. At rest, geometry and gravity hold the keel in place.

Thirdly, again inspired by the Herreshoff's, father and son, we framed the *Alerion Class Sloop*'s deck longitudinally. We glued light stringers (1 inch molded by ⅝ inch sided) to her ½ inch plywood deck plating and laid it over her interior bulkheads. At the partners and cockpit opening we bent into place and fastened "deck beams," 5 in all, of the same stock. We installed spacer blocks in between the stringers to tie the deck plating and beams together to form vierendeel trusses. The deck frame was entirely made of linear pieces, its double curvature bent, not cut. It was light, strong, and easy to make.

These three innovations gave us a structure superior to the industry standard (which was not very standardized at that point) and we were able to build *Alerion Class Sloop*'s in a little over 1,100 man-hours, about a third of the norm for custom plank-on-frame boats. The boats had the grace, performance, and beauty of Herreshoff's original. Without metal fasteners or frames, their hulls were lighter, allowing them to carry more ballast and so sail better. They were also easier to care for and were far more durable. Four decades later, all are still sailing and most are in like-new condition.

But the job of developing techniques for epoxy-bonded boatbuilding was not over. The *Alerion Class Sloop* was a simple, inshore day-sailor. Much of her economy came from building the same boat over and over. The job remained to learn how to extend what we had learned building the *Alerion Class Sloop* to the building of one-off oceangoing boats. I did not know then, but, once again, it was going to take three tries to get it all just right.

FANCY

In 1983, Sanford Boat Co. organized an R & D partnership to develop techniques for building a 53 foot design named *FANCY*. She was built from 1984 to 1985 at Sanford-Wood Marine, Inc., in Richmond, California.[3]

We sought three techniques: a method to spile the planking economically for a one-off, a method to build the keel/centerline structure construction that would incorporate the tankage and provide foundations for the mechanicals required for an offshore boat, and a method to install the pre-built interior economically.

This last, the economical installation of the interior, is difficult. As we learned with the *Alerion Class Sloop*, the completed hull shell, when

---

3. In 2001 Rick Wood and I sold Sanford Wood, which is now operated as Keefe Kaplan Maritime.

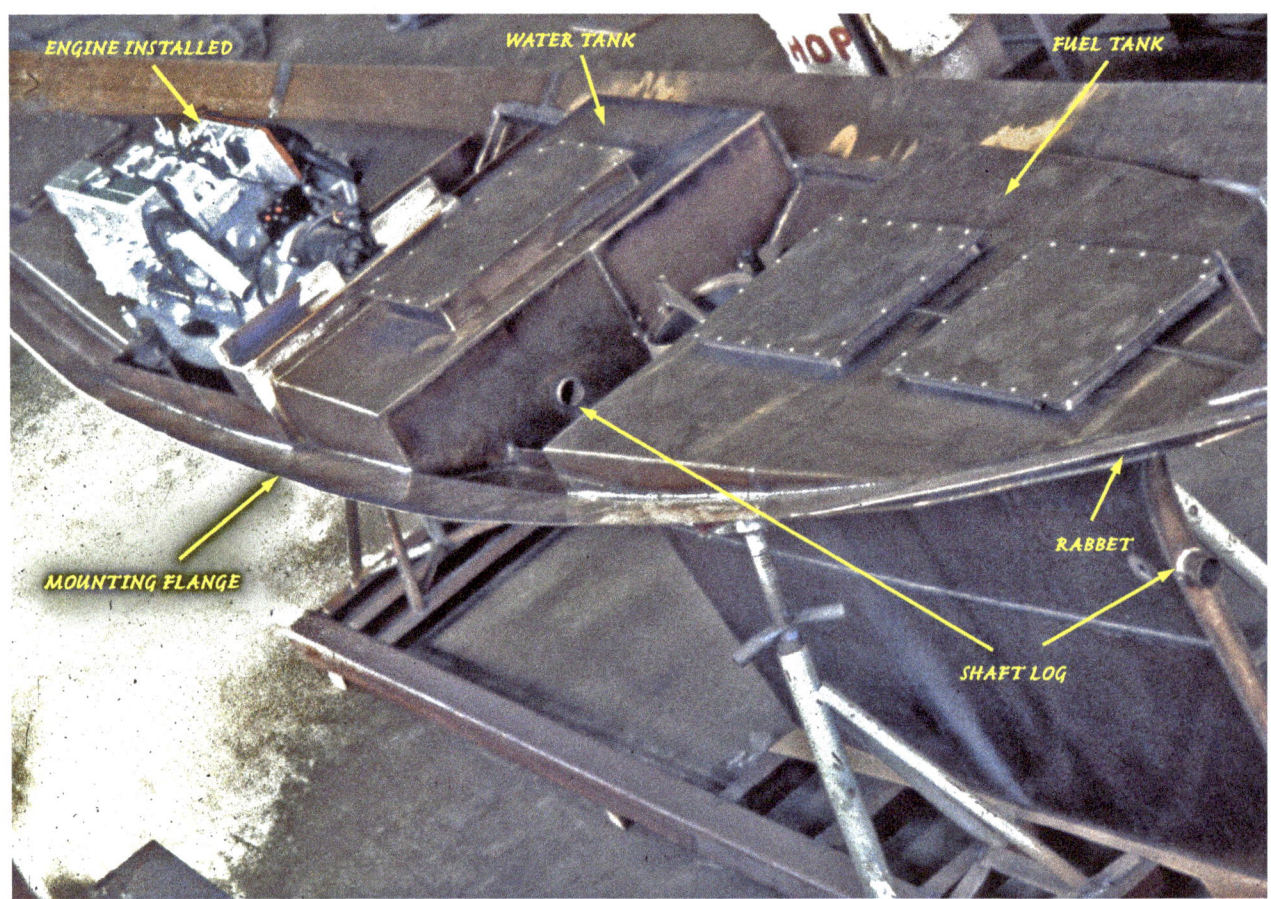

3-7 *FANCY*'s keel. The keel was built first, of copper-nickel, an alloy of 90% copper and 10% nickel. The red metal is mildly antifouling and very stable in a saltwater environment.

upright, is a bowl in which it is difficult to work. The bowl shape restricts movement in and out, of people, tools, and material. It is hard to keep clean, and until the sole is built, it is hard to even stand up in. It is also difficult to find precise locations and orientations for the parts being fitted into it. And each of the many interactions between the hull shell and the accommodation pieces must be precisely fitted. Cheaper practice is to make rough fits and then conceal the joints, but this is unseaworthy (**RULE 3**).

We explored these issues by building *FANCY* as follows:

Firstly, we built her keel of a copper-nickel sheet metal weldment. The lead ballast was poured with straps that were welded to the copper-nickel. This was analogous to the lead epoxy keel of the *Alerion Class Sloop*. The

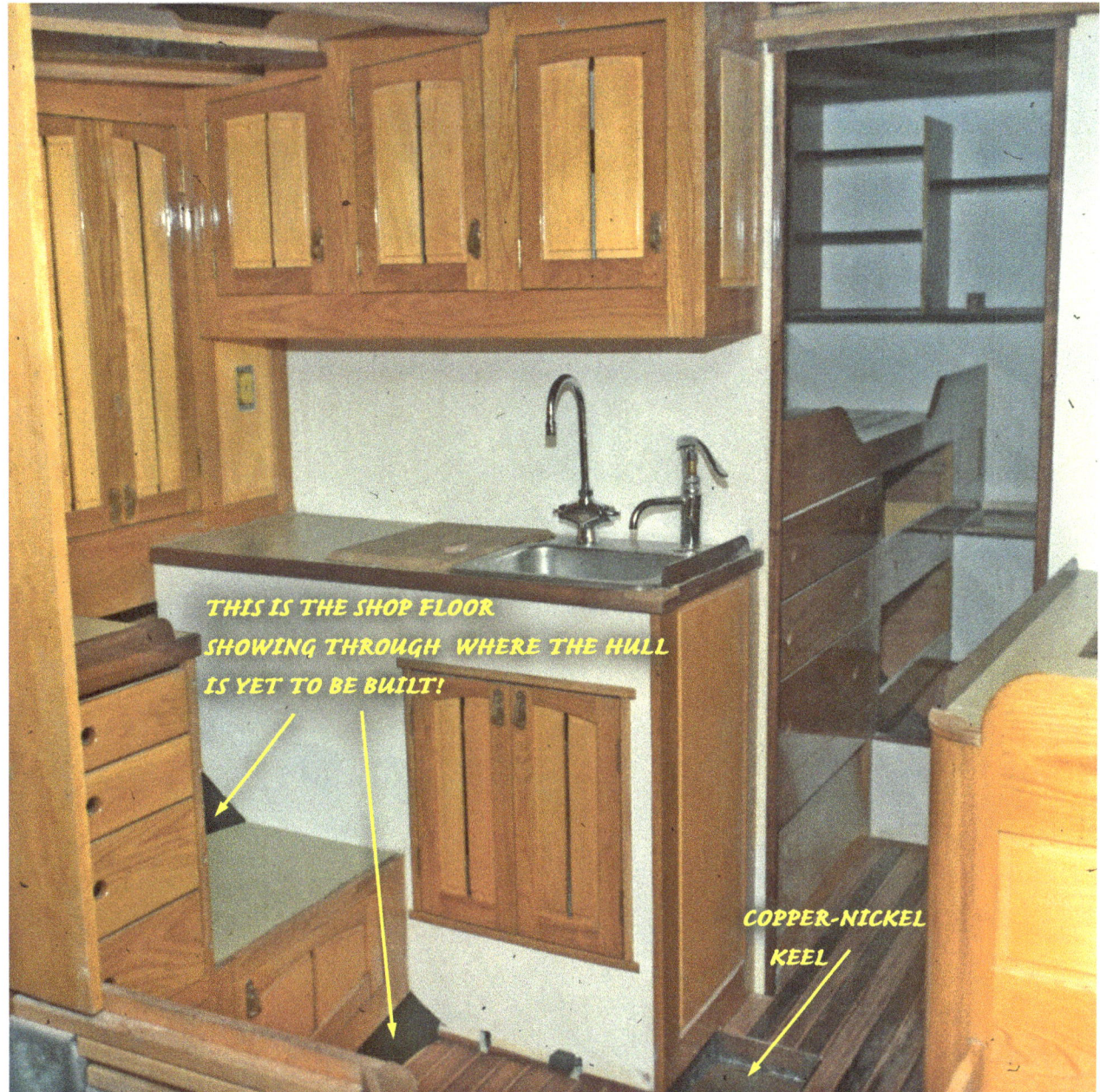

3-8 **FANCY's mold.** Here FANCY's interior is set up on the keel and held together by the deck. The photo is taken from the main saloon, and the house has yet to be built. You can see through the opening for the galley stove to the shop floor as the hull has yet to be built.

The interior is quite complete, with the fo'c'stle sole in place and one of the galley foot pumps installed. The interior will be used to mold the hull in the next step of her construction.

copper-nickel sheet (⅛ inch and ¼ inch thick) formed the bilge and mast step and incorporated integral tanks, engine foundations, and the shaft log. *FANCY* was to be a keel boat. so she was spared the complication of centerboard box and her rudder was hung separately on a skeg (also copper-nickel.)

Secondly, after the keel was completed right-side-up on the shop floor, the stem, horn timber, and transom were attached to it. Then the entire interior accommodation was set up on the keel. Over the interior, the deck, which had been pre-built, was placed, fixing the tops of the bulkheads in place and tying the whole complex together. At this stage the mechanicals were roughed in.

Finally, with the inside and deck complete the hull was built around what had, up till then, been an open structure. Since the interior accommodation was the hull mold, there was no issue of fit, nor did it matter if certain pieces were slightly out of alignment. Generally they were not, because working in the

3-9 *FANCY*. Sailing on San Francisco Bay soon after her launch in 1985.

open, it was easy to get everything in proper place. Running mechanicals was also easy but required care lest something be forgotten, or worse, installed in such a way that it would lack proper access later (**RULE 3**).

We lined off the hull traditionally by creating a taper function for the planks and marking their intersections on six of the bulkheads. We then took a 55-foot batten, made of 1½ inch steel angle, and set it up on the structure adjacent to the marks. The batten exposed any unfairness and allowed us to correct it at this stage. Next, using a router that ran along the batten we cut notches in the accommodation pieces wherever they touched the hull surface. Into those notches we glued seam battens, (sided 1½ inch, molded ¾ inch). Once all the battens were in place, we laid our inner plank stock, ½ inch thick, against the battens and directly marked the spiling onto the stock from the battens. Similar to the *Alerion Class Sloop*, we cut inner and outer planks in one shot. We glued up the hull similarly and, when done, the boat was ready for paint, hardware, and rigging.

Our three gambits tried with *FANCY* worked out pretty much as expected, but we learned a few things. The copper-nickel bilge, machinery foundations, and tankage were wonderful, but the metal work required special tools and skills. Building a metal boat is no harder than building a wooden one, but it requires completely different skills and tools, so *FANCY* really required two separate boat shops. Such is rarely practical.

Using the interior accommodation for the mold did eliminate fit as an issue with building the interior. *FANCY* had an elaborate interior, and it went together easily in the open on the shop floor. But using it as the mold had two drawbacks. First, the fine woodwork of the interior took some abuse from the rough work of building the hull. And, second, it was necessary, using this technique, to build the hull right side up, a drawback with gluing the planking and finishing the exterior.

The seam battens made spiling the planks easy enough that fore and aft planking was as economical as the more common diagonal. But we discovered the horizontal seam battens created water and dirt traps inside the hull. Sailing her, we learned that there should be no impediment to the free flow of interior water and debris to the bilge.

FANCY was a very fine boat, simple and strong. She was pretty and sailed fast; she was dry, with a gentle precise helm. She sails in New England today. The drawbacks to her construction though were significant. We needed better techniques for the future.

STARRY NIGHT

The future was to be some time on because, after *FANCY*, I took a break from boatbuilding.

I spent the next 25 years working with real estate, all the while cruising on my *IMPALA* a couple of months a year and crossing oceans to get to the cruising grounds. I spent some time restarting *Alerion Class Sloop* construction on a limited scale. But the call of boatbuilding is strong and, in 2008, I began a project to continue experimenting with cold-molding/epoxy bonding techniques by building a 40 foot oceangoing cruising boat to be named *STARRY NIGHT*. She was built 2008–2009 by Brad and Mike Pease at their Boat Works and Marine Railway in Chatham, Massachusetts.

By 2008, cold-molding was no longer radical. Although it was not well understood by the public, it had become an accepted technique in the industry. Much had developed since *FANCY*'s launch. A simplified shell construction had been standardized, which used first a thick strip-planked layer over plywood molds to establish the hull surface. To this first layer, diagonal veneers were laminated and held in place by vacuum bagging. Vacuum bagging places a plastic film (the bag) over a glued-up layer of a cold-molding. The air is sucked from between the bag and the glued veneers so that atmospheric pressure

3-10 ***STARRY NIGHT*'s mold.** The mold elements were laser-cut from digital files. Notice the jigsaw puzzle joint used where a mold element was bigger than a sheet of plywood. The laser can also mark layout lines or instructions directly on the wood parts, though we did not do so with *STARRY NIGHT*.

3-11 **Turned over, mold stripped.** *STARRY NIGHT*'s completed hull off the mold and uprighted. The men are coating her inner surface with sealer. Notice the engine beds molded directly into the hull.

3-12 *STARRY NIGHT* fo'c'stle. The forward cabin and forepeak, complete, just before the deck is assembled.

squeezes them tightly to the underlying layer while the glue sets. The bag does a much better job than the staples it replaces, eliminating any voids in the layup. With this layup technique, the unpleasant look of the exterior diagonal layer was dealt with by painting the hulls.

Another major development since *FANCY*'s build was the introduction of inexpensive computerized lofting, which had become available even to small shops. Matt Smith, a brilliant young Ted Hood trained naval architect from Barrington, Rhode Island, had been lofting for Pease for some time using the new technique and he became deeply involved in the *STARRY NIGHT* project. Computerized lofting combined with computer-

3-13 **A wooden boat is a beautiful thing...**

controlled laser cutting of sheet materials makes the construction of building molds and other flat work almost trivial. Further, the ability of the computer to develop true shapes allows the spiling of carvel planking in seconds rather than the days it took doing it the old way. Computerized lofting and cutting is still new and the technology asks the boatbuilder to find additional uses for it. Matt Smith is a leader in the development of the cutting techniques. And Laurie McGowan, an extremely inventive yacht designer who lives a half hour away from Slocum's birthplace in Nova Scotia, has developed computer generated spiling.

STARRY NIGHT was built using vacuum bagging and computerized lofting, but she also used a novel technique of hull construction. *STARRY NIGHT* had a stringer-generated hull, maybe the first of its kind (see chapter 7). Rather than her hull being generated the conventional way, with close spaced frames shaped over molds, her shape was generated by longitudinal stringers. Her mold held in place a shear clamp and five stringers each side. Over them, as secondary members, were bent light frames. Then the planking was attached to the frames. The frames touched only the stringers and the planking, so they needed no beveling nor did they require precise location. The interior accommodation rested only on the stringers and never came in direct contact with the hull plating or the frames.

Stringer structure has a long history in metal construction and in aircraft construction. For sailboats, it was pioneered by Francis Herreshoff in the 1920's, perhaps inspired by his work with Starling Burgess, an early aircraft builder. It has a number of subtle structural and practical advantages, which will be discussed later in this book.

Being a cruising boat, *STARRY NIGHT* had an elaborate interior. Her accommodation was mostly prefabricated flat panels made of butternut plywood and butternut solid stock. Thirty years after inventing the *Alerion Class Sloop*, I had yet to discover a method for quickly installing such an interior into a finished hull, so in reality we took a step backward from *FANCY*. After the hull was uprighted and before the deck was added, we found our way within the bowl of the upturned hull and installed the sole. Then we located and installed key bulkheads. This fussy process was easier than it had been with the *Alerion Class Sloop*, because only the bulkheads touched the stringers and they touched only at points in rather long junctions with the hull shell.

To make the bulkhead/stringer joint, we invented a new piece, which I named the **bulkhead block**. The bulkhead blocks were oak, a little narrower than the stringer (3 inches) and 1 inch square in section. Their bottom surface is beveled to the angle of the stringer with the bulkhead. The bulkhead blocks were glued and screwed to the stringers and the bulkheads were bolted to them. A typical bulkhead would have five or six blocks, making a strong (and removable) joint.

But we still had to find our way within the bowl; we did not know precisely where the blocks should go. It took time and effort to get the first pieces correctly located and oriented, else the entire interior would become a mess. There was more thinking to do. Easy location of the bulkheads became a goal for the future. I leave description of the discovery of that technique to chapter 9.

Building the boat over stringers resulted in a fair, easily built, and very strong hull (**RULE 1**). As we will discuss in chapter 7, it resulted in a hull of superior strength-to-weight ratio compared with the common techniques used for wood, metal, or plastic construction. We obtained an effectively thick hull with visual access (**RULE 3**).

There was also more thinking to do about the hull/deck joint. As with both *FANCY* and the *Alerion Class Sloop*, we joined hull and deck by building up the shear clamp so that it was wide enough to provide sufficient glue surface for the structural deck. This meant using a lot of wood. The shear clamps were hard to bend and after installation they required the cutting of a continuously varying bevel along the top to fit the varying angle between hull and deck. This bevel was hand-cut, in place, and involved quite a bit of skillful hand labor. It was an expensive joint.

There were other, more minor issues. But for now, it is enough to say that *STARRY NIGHT* turned out as a strong and stable boat, easy to sail, comfortable, fast, with a very cozy cabin (**RULE 2**). She obeyed all the rules.

Now it is time to digress into some of the fundamental technical issues of sailboats. There is science as well as art in boatbuilding and the one cannot stand without the other. Chapters 4, 5, and 6 will discuss structural forms, a sailboat's loadings, and wood as a material to build a sailboat with. Then we will come back to actual building in chapter 7.

PART TWO

Science

Yet that dim light of my compass on this dark Antarctic night made me look with tenderness on these wrought planks, flesh of the beautiful trees of my country, fashioned by human knowledge into a boat.

—Vito Dumas, writing as he approached the
Drake Passage, *Alone through the Roaring Forties*

CHAPTER 4

Sailboat Structure

STRENGTH OF MATERIALS is the study of how loads effect a structural element and whether structural elements are strong enough. Structural analysis has been with us since the Greeks, mostly in the form of rules of thumb. In 1757 Swiss mathematician Leonard Euler developed an elegant theory to determine the buckling load of a thin column. At the beginning of the 20th century, Stephen Timoshenko established strength of materials as a science by setting forth comprehensive theories for determining the stresses in structural elements and the deflections those stresses caused.

Timoshenko's formulas have two components. The first component concerns geometry--the size and shape of the structural element and the size and shape of the loading it carries. Geometry shows some surprising facts, particularly that **thickness**, is the most important dimension in determining strength. Thickness determines the effectiveness of beams, columns and shells. It controls buckling which is a major and non-intuitive form of structural failure.

The second component of Timoshenko's formulas is the inherent strength of the material being used.[1] The strength is quantified by numbers known as **mechanical constants**, and each particular material has its own. The most important mechanical constants for our purposes are modulus of elasticity, symbolized "E," which is a measure of stiffness of a particular material, and maximum strength, symbolized "σ_{max}," a measure of when a particular material will break. The mechanical constants for common boatbuilding materials are tabulated in table 6-1.

With sailboats we are mostly concerned with the bending of beams, the compression of columns, the stretch of tensile members, and the deformation of shells. The hull of a sailboat is itself a beam; her mast is a column; her rigging parts are tensile members; her hull skin, a shell. We'll look at each in a little detail.

Beams

The **beam** is a simple structure. You lay a plank over two sawhorses, and it spans the space in between. Your plank will hold up heavy things. The plank is a beam, the heavy things are its load, and the horses are its supports.

Beams carry loads directed perpendicular to their length. In our plank's case, the weights push down vertically. The plank transforms their vertical push into horizontal loads that run along the length of the plank. The upper part of the plank has a compression load and the lower part a tension load. I won't go into the theory of moments, which Timoshenko used to obtain his formulas, but I will point out that besides affecting how much material there is to resist the load, the depth of the beam also determines the separation between its compression and tension forces. Thus the depth dimension affects the amount of load the beam can carry twice and, indeed, depth shows up as a squared factor in the formulas.

The forces may be better visualized by looking at a **truss,** which is a beam reduced to its elements. In a truss all the forces are aligned with its elements, which are in either pure tension or pure compression—there is no bending. Looking at a truss, one can visualize the forces hidden within a beam, compression on the top chord and tension on the bottom. The diagonal members and the vertical members carry the horizontal and vertical shear that join the compression of the top with tension of the bottom. Although these shearing forces are responsible for the hogging of old wood boats whose fastenings have failed, they generally affect sailboat structure only in minor ways. We acknowledge them but will not study them further.

---

1. Here "strength of the material" means how strong some material is (e.g., steel). It sounds just like the phrase "strength of materials," which means the theory of how strong structures are. This is an example of nearly the same words being used to describe two quite different ideas. It is poor and confusing terminology. Henceforth, I will use "structural analysis" for the more formally proper "strength of materials."

CHAPTER 4 — SAILBOAT STRUCTURE

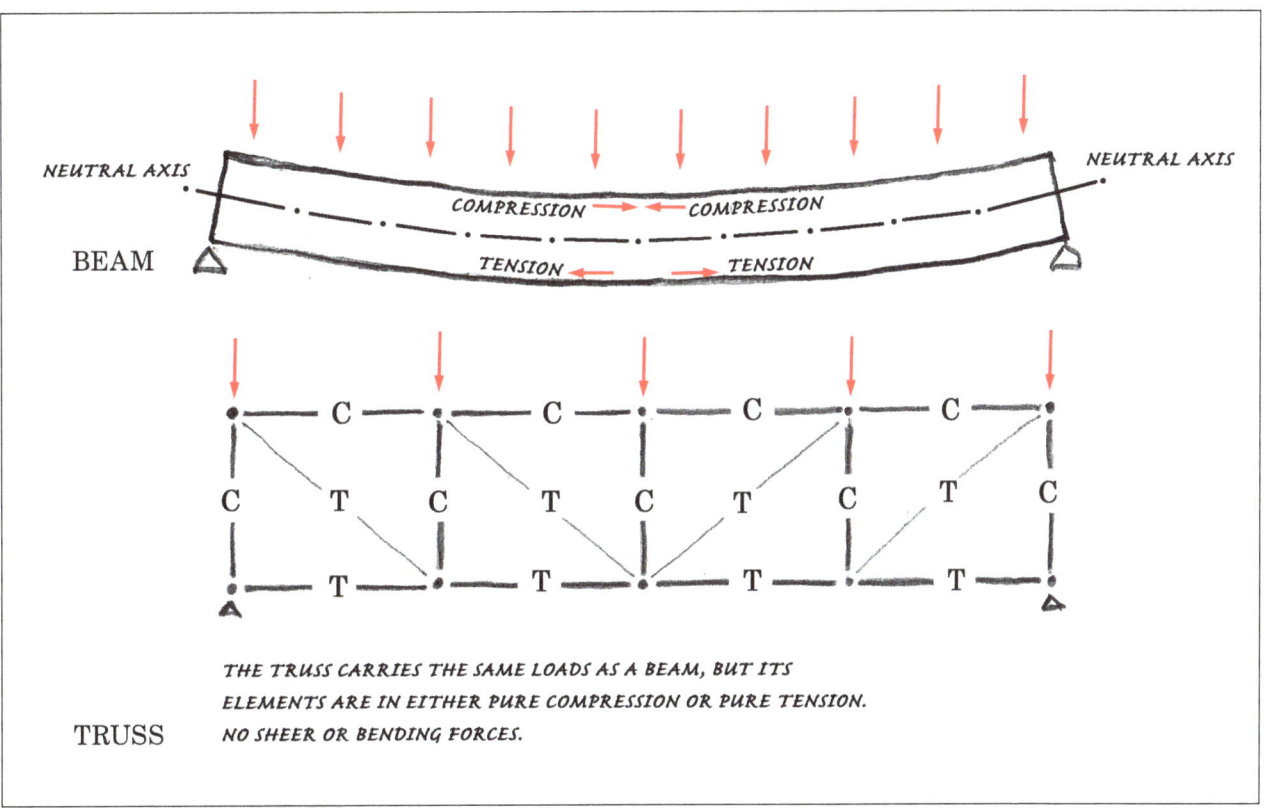

4-1 Beams. Beams carry loads *perpendicular* to their length. The beam changes the perpendicular load into compression and tension stresses *along* its length. The most important dimension for a beam is its depth or thickness in the direction of the load. The truss illustrates the forces internal to the beam. A truss's elements carry only tension or compression; they have no bending or shear loading and the joints are hinges.

The tension and compression stresses reduce towards the middle of the beam. At the middle the stress is zero and the line of zero stress is called the "neutral axis." The material near the neutral axis is not working very hard, so beams may be made more efficient by removing some of that material, saving a lot of weight with only a little reduction in strength. This results in the I-beam where the center material is cut away leaving a thin web just thick enough to carry the shear loads. Another way to reduce the central material is to replace the central material with a lightweight substitute. This makes a sandwich construction, where a lightweight core, often foam, separates strong skins. The core holds the skins apart, achieving the necessary thickness of structure but only adding a little weight. The core need only be strong enough to carry the shear loads.

Timoshenko's science allows us to calculate how much load a beam can carry and how far it will deflect. We need not go into the

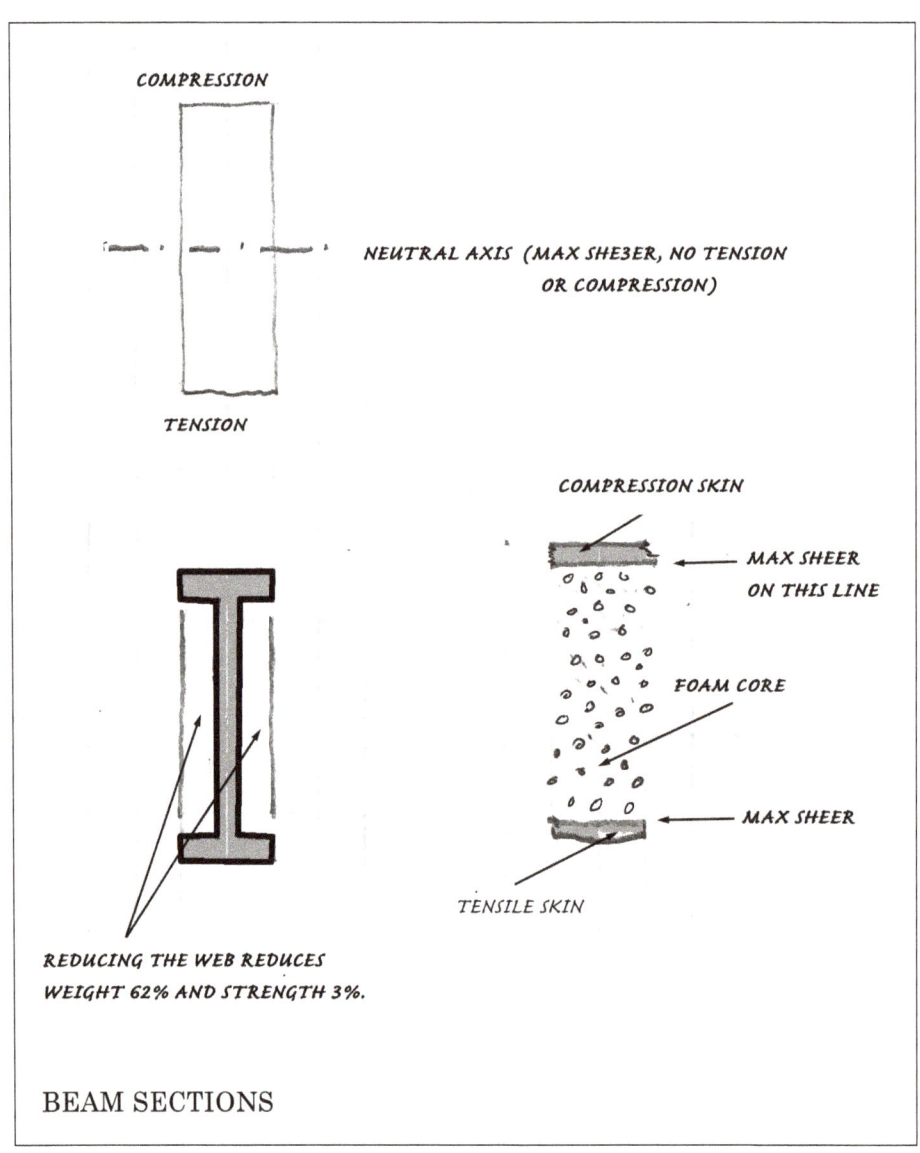

4-2 **I-beam and sandwich.** Since there is little stress near the neutral axis of a beam, material can be removed from the center, or weaker, lighter material may be substituted. Either will reduce weight and increase efficiency of the beam.

details here, but the important fact to remember is the *geometrical* one.

RULE OF BEAMS: When the depth of the beam is doubled, it becomes four times stronger and eight times stiffer.

Such is not so if the width is doubled.[2] That doubling merely doubles the strength and cuts the deflection by half. It also doubles the weight. The effect of increased thickness is important. By putting material in the right place, by using the right geometry, we can increase the efficiency of our structure.

Columns

Columns support compression loads along their length. Columns are characterized by their slenderness ratio, the ratio of their thickness to their length. Short stubby columns fail from the crushing of their material; it takes a big load to do it. Slender columns fail by buckling, long before crushing occurs. Buckling occurs when some small eccentricity starts the column bending sideways "out of column." If the load is big enough, the load, now off center, generates additional moment that bends the column farther out of column. The effect feeds on itself and the column fails catastrophically.

The load required to buckle a column is called the **Euler load** after Leonard Euler, the man who developed the theory and formula for calculating it. As might be expected, because bending is involved, he discovered that the critical load was dependent on the square of the slenderness ratio.

RULE OF COLUMNS: Twice the diameter, eight times the weight and eight times the strength.

At a certain (low) slenderness ratio, the Euler load becomes equal to the crushing load and the column becomes stubby. The fixity of the ends of a column, whether they are restricted from rotating, or moving sideways, changes the effective length of the column and hence its slenderness ratio and load-carrying capacity.

Tensile structures

Tensile structures carry tension and axial load along their length. Unlike compression structures, tension structures are inherently stable. Their carrying capacity is proportional to the cross-sectional area and they require no support to stay in place. The thickness of the element has no particular bearing on its strength. With tension, there is no analogue to compression's buckling, which is one of life's interesting asymmetries.

---

2. The distinction between width and depth of the beam is determined by the direction of the load. The depth is parallel to the load direction, whereas the width is perpendicular to it.

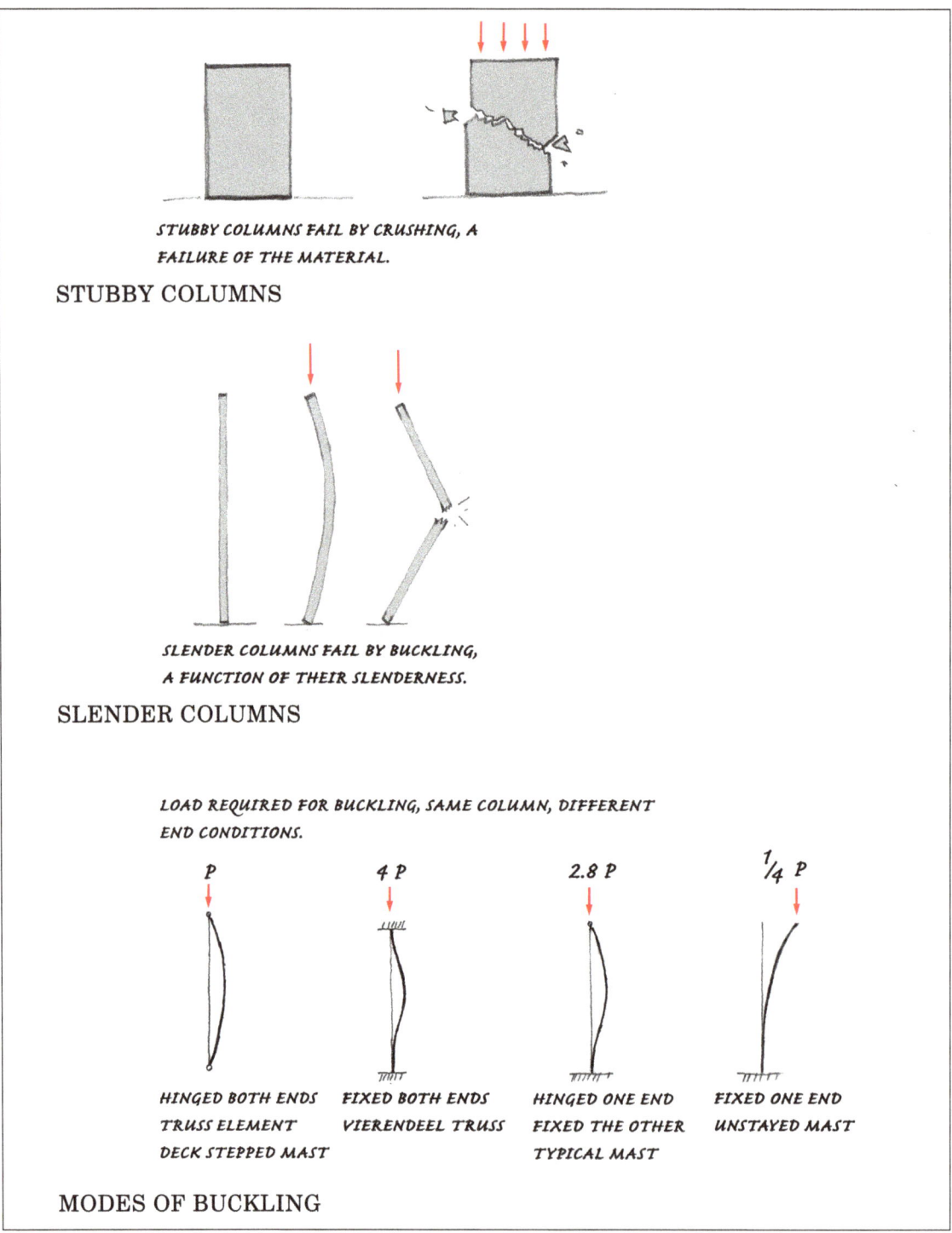

4-3 Columns. Columns support compression loads along their length. Stubby columns fail only at loads that crush their material. But slender columns fail by buckling. The slightest eccentricity will cause them to bend. At a certain loading the bend itself will generate a bending moment that bends them further. Then they will collapse suddenly. The critical load was first determined by Euler in 1757. Notice that the support conditions at the ends of the column determine the effective length of the column and hence its effective slenderness.

A tension structure may act as a beam carrying loads perpendicular to its span if it has sag. It will support the load with tension forces along its length. A common example is the suspension bridge. The tension generated in the wire is determined by the amount of sag (the more sag, the less tension).[3] The headstay of a sailboat carries the sideways load of the genoa the same way.

Vierendeel truss

There is a particular form of truss, called a vierendeel truss after the Belgian Arthur Vierendeel, who popularized the form in 1874. The vierendeel truss has only vertical web members—no diagonals—which makes it useful because it is easy to move through. To carry load, the vierendeel truss depends on rigid joints and the stiffness of its elements. If the elements are flimsy, they will buckle, and the truss will fail at low loads. But if the elements are stubby, a vierendeel truss can be very strong. We will discuss it later in reference to hull shells.

The vierendeel truss is not really a truss. The members of a true truss carry no bending loads and the joints are hinges. Quite the opposite, the elements of a vierendeel truss do carry bending; that is how they avoid the otherwise necessary diagonals. Furthermore, the joints of a vierendeel truss must be rigid. If they were hinged, the vierendeel truss would collapse of its own weight. But we will call it a truss because that is common practice and it sort of looks like a truss.

Shells

A shell is a thin, curved surface that supports loads perpendicular to its surface. As with an arch, the geometry of the shell redirects the loads that are perpendicular to its surface into compression loads that run parallel within its surface. The material of the shell resists this compression very efficiently as long as it is thick enough and has sufficient curvature. Curvature adds effective thickness beyond its actual thickness. A shell fails by buckling when it has too little curvature, or too little thickness. As long as the shell maintains its curvature. it is very strong—the more curvature, the stronger. The need for curvature leaves a shell vulnerable to local deformations that might be caused by some local loading. If the shell is flattened by a local load, it will buckle and break.

Shells can be characterized by the ratio of their curvature to their thickness. At a low enough curvature/thickness ratio, they act as stubby columns and are extremely strong. An

---

3. "Sag" is not really an engineering term, but I use it because the tension structure we are most interested in is the genoa-loaded headstay. Deflection of the headstay from straight is called sag.

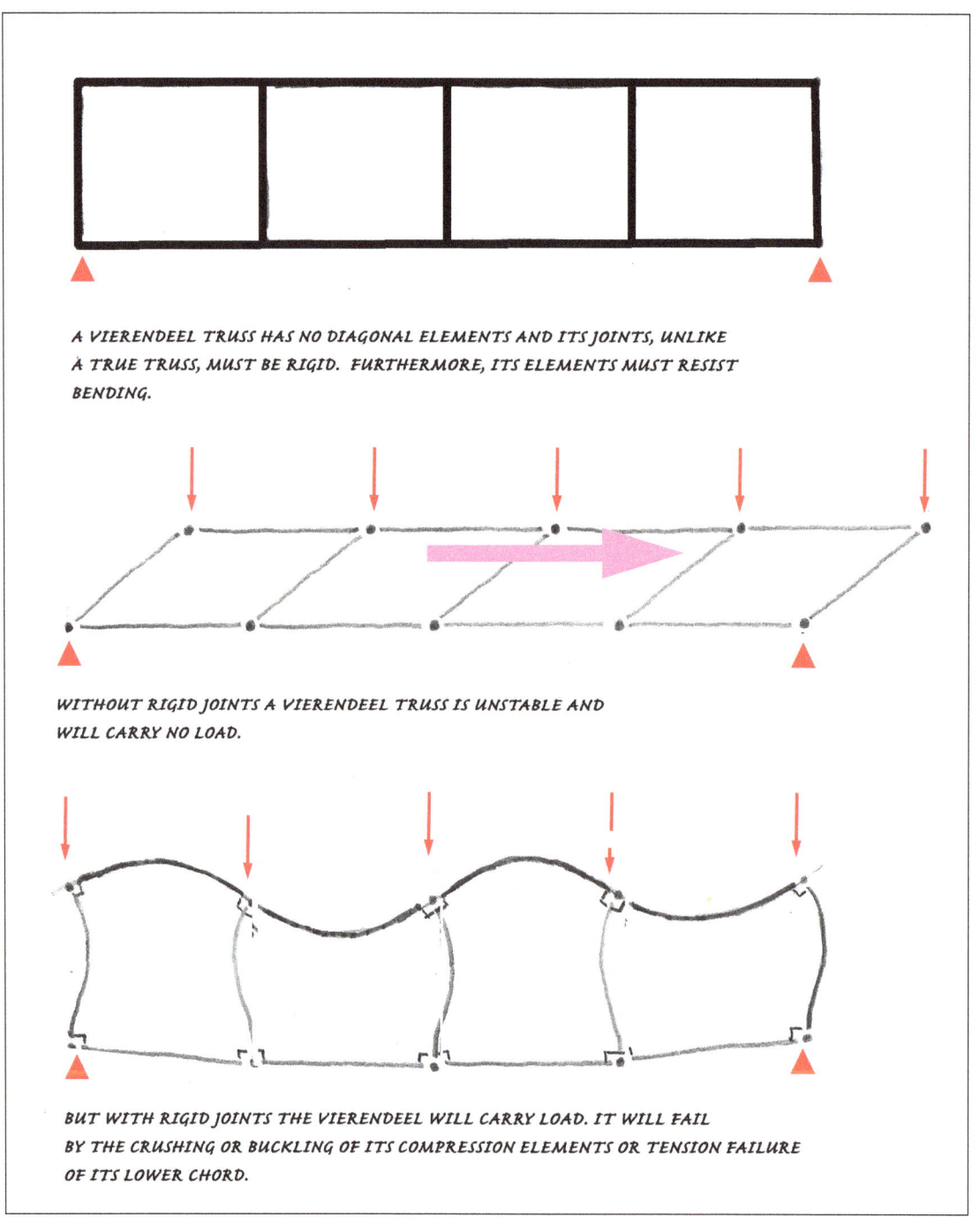

4-4 Vierendeel. The vierendeel truss is not really a truss. A truss has hinged joints; a vierendeel truss does not. A vierendeel truss with stubby members can carry large loads, failing only when the top chord or the verticals are crushed or the bottom chord is broken in tension.

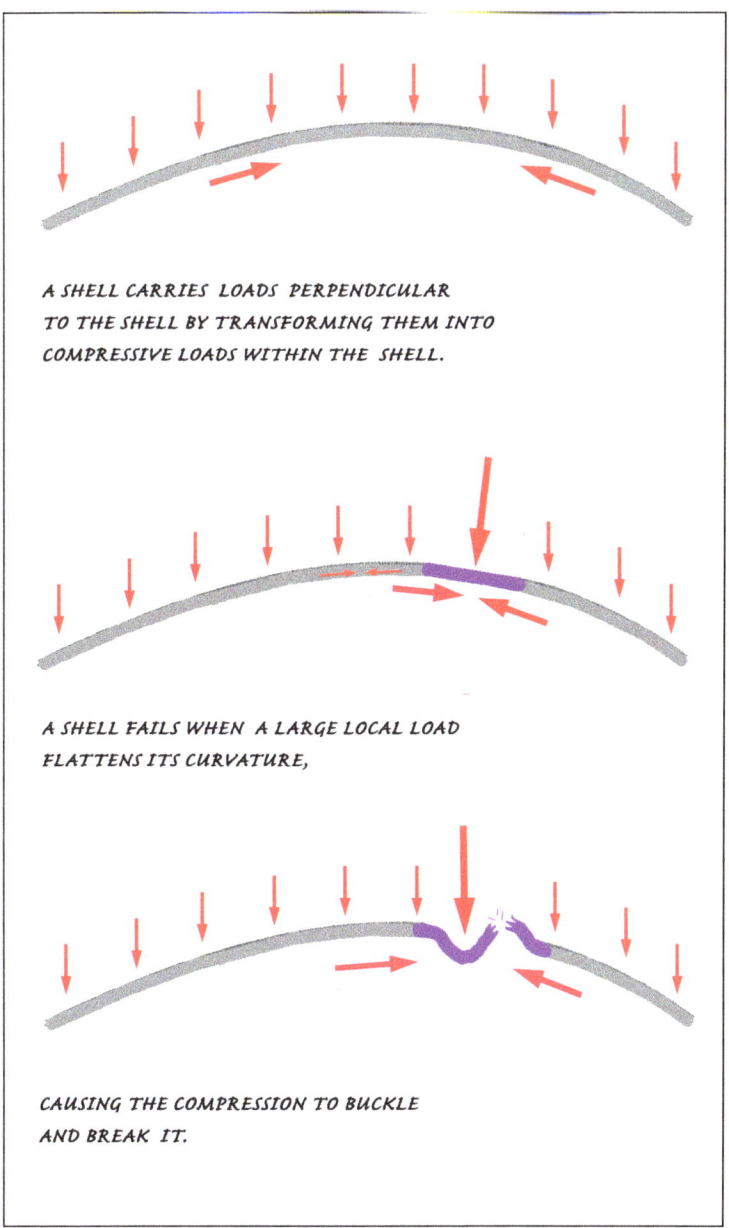

4-5 Structural shells. Shells are strong as long as their geometry is intact. When deformed and flattened, the shell becomes "thin"; it buckles and fails.

eggshell is doubly curved with a curvature/thickness ratio varying between 40 and 100. Although eggshell, as a material, is weak, even a strong person cannot break an egg by squeezing it. On the other hand, an aluminum soda can, which has a curvature/thickness ratio of about 280, is easily crushable by hand. Shells fail in buckling, which brings us to the following:

RULE OF SHELLS: A shell twice as thick is four times as strong.

We will come back to this subject towards the end of the chapter when we discuss shell thickness and ask how much thickness is enough.

Sailboat loadings

To design adequate structures, one not only needs to know about the structure, one also needs to know about its loadings. The loads acting on a sailboat, when she is underway, are complex and mostly indeterminate. While our computers do not, as yet, accurately model a sailboat in a seaway, we do have a few thousand years of experience with what is and is not strong enough. This experience has been digested into rules of thumb, which are sufficient for practice. Called **scantling rules**, they offer sizing for the elements of a sailboat's structure. Lloyds of London and Nat Herreshoff both published scantling rules,[4] and there are others. Both rules have created successful boats. Both are closely tied to the specific forms of construction for which they were devised. In my work, I usually begin with Herreshoff's rules. They are well tested by time and produce a stronger, lighter boat than Lloyds' rules. But cold-molded construction is different enough that we must move beyond Herreshoff and develop his rules into new ones. Let's start with what we do know about loads on sailboats.

Gravity loads

The two most basic forces on a vessel derive from gravity, the buoyant force of water pushing her up and the weighty force of gravity pulling her down. Buoyancy pushes perpendicular to the hull surface with a force dependent on depth below the surface, roughly ½ pound per square inch per foot of immersion. Gravity pulls down vertically on the vessel with a force equal to her weight.

Archimedes tells us the two forces are equal. But they are not uniformly distributed along the length of the boat. The unequal distribution causes a loading on the hull. With a normally shaped hull, the force of buoyancy is concentrated amidships while the force of gravity is not. Thus, a vessel's hull acts as a beam, supporting its ends cantilevered out from central support.

---

4. Lloyds Register of Shipping, *Rules and Regulations for the Construction and Classification of Wood and Composite Yachts*. For Herreshoff, see Kinney, *Skene's Elements of Yacht Design*, pp. 243–261.

A. STILL WATER

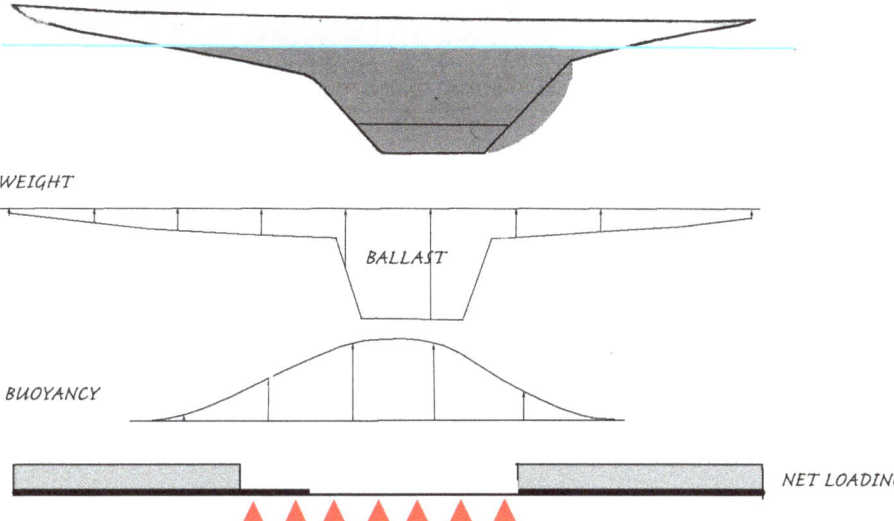

B. CREST AMIDSHIPS

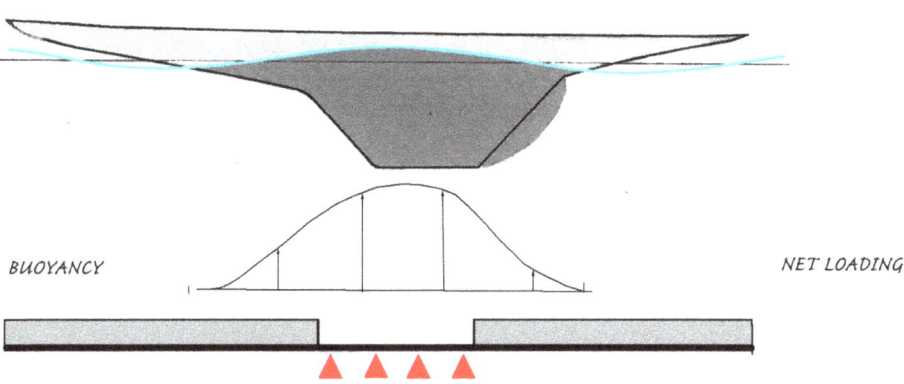

C. TROUGH AMIDSHIPS

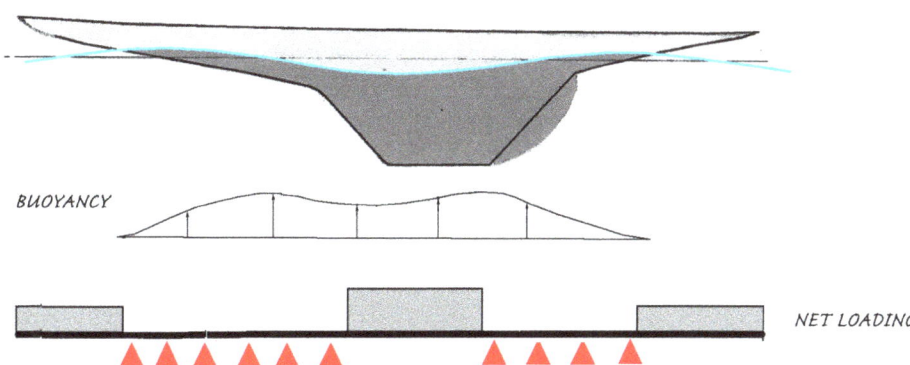

4-6 Gravity loads on a sailboat. The hull chosen is NGH's fabulous *RELIANCE*, whose extreme hull was mostly only a beam. While she had some room for sail stowage, most of the interior was occupied by a steel space frame reminding one of Eiffel's tower. Normal boat hulls act, along with their decks, as tubes, which make strong beams.

Her deck is stretched in tension; her underbody is squeezed in compression. In a seaway, with a wave in the middle, this loading is significantly increased.

Or it is reversed. With a wave at either end, the hull is supported at the bow and stern with its load in the middle. The deck undergoes compression and bottom tension, just the opposite of the smooth water condition. A boat running in a seaway will have the constantly alternating loading condition of bottom and deck alternately loaded in compression and tension.

The magnitude of the tension and compression loads is proportional to the vessels' displacement (weight) and the vertical dimension of her hull. In a conventional design the hull is not particularly slender and the weight of the ballast, a significant portion of the total displacement, is also concentrated amidships, so the loads generated by buoyancy and gravity are fairly low.

Nonetheless, one occasionally sees a boat, usually a plank-on-frame built wooden boat, with sagging ends. Such a boat is called "hogged." Usually, with fastenings failing, she can no longer hold herself up. She is near her end.

Wind loads

Sailing, a boat feels the force of the wind interacting with her sails. The wind creates loads that are transmitted to her hull by her spars and rigging, which are her customary working loads, encountered frequently, over long periods of time. Running before the wind, the rigging loads are moderate, but as a vessel comes closer to the wind, the situation changes. Large loads are imposed on the boat fore and aft along her centerline and athwartships at her mast.

Athwartships, the force of the wind interacting with the sails causes the boat to heel. Her ballast keel opposes the heeling forces. The result is the tension on her weather shrouds, which is applied to the hull at the chainplates and runs through the hull shell down to the mast step. The mast is compressed by the shrouds pulling down and the mast step pushing up. Secondary forces squeeze the deck at the mast partners. In a sailboat of normal beam and draft, the magnitude of the shroud load is approximately equal to the displacement of the vessel.

Fore & aft, the hull is loaded by the headstay and backstay/main sheet. The headstay transforms the wind's sideways push on the jib into a large tension on the headstay. The amount of tension depends on the sag of the headstay and the size of the jib—with a genoa in racing trim, it is a large amount. The headstay pulls the bow up and pulls the mast down; the main sheet and/or backstay resists the forward pull of the forestay. They pull the aft end of the boat up. The hull acts as a beam supported at the ends and with a large point load (the mast) in the middle. This loading induces compressive loads in the deck and tensile loads in the bottom.

Theoretically, the headstay loading can approach the infinite, depending on how straight one wishes the headstay. In practice it is limited by the flexibility of the mast, the headstay, and the hull. Their strains relieve the loading by allowing the headstay to sag. In practice the load on the headstay is maybe a quarter of the boat's displacement.

Both the fore & aft and athwartship wind loads put large tensile stresses in the hull. The fore & aft stresses want most of the hull material running fore & aft, and the athwartship loading requires strong athwartship hoop strength in the region of the mast (**RULE 5**). Traditional wood construction carries the athwartship load with the frames and their floors. Cold-molded hulls carry it within the hull laminate.

Dynamic loads

Dynamic loads are imposed on a vessel when she is suddenly accelerated (in practice generally, *decelerated* by crashing into something). Dynamic loads are mostly indeterminate; their magnitude is difficult to calculate. Some dynamic loads, those imposed by wind gusts and waves, are ubiquitous and must be dealt with regularly. They are accounted for by using a factor of safety. Dynamic loadings imposed by collisions, groundings, and knockdowns are rare, but extreme. One way to approximate them is to argue that a five G acceleration is the maximum the crew could survive, making it a practical limit for design.

Knockdown loads are a bit special. A knockdown is not to be confused with a broach. A broach, which can capsize an unstable boat, is caused by an over-canvassed boat with poor steering qualities, getting out of control and rounding up into the wind—often with a spinnaker flying. People with sensible boats don't experience broaches. They were uncommon before the IOR rule brought big dinghies to the starting line. Waves help but are not required, and dinghy shapes will broach in as little wind as force 5.

A knockdown, on the other hand, is caused by large waves in gale conditions. In certain areas, large irregular seas occur—think of the Drake Passage, the Gulf Stream, or George's Bank. With cross seas or shoal water, they can become steep, unstable, plunging breakers that will lift a boat up, roll her over, and drop her into their trough which might be 20 feet below. When the vessel hits the trough, from that height, the water is quite hard. The impact puts extreme lateral loads on the mast and the **leeward** cabin coamings. Usually all gear is swept off the decks, the mast breaks, and, on a traditionally built wood yacht, the lee cabin coaming splits, down flooding the hull.

House coamings of a traditionally built wood boat are made of a plank of wood running parallel to the deck. In the vertical direction, they

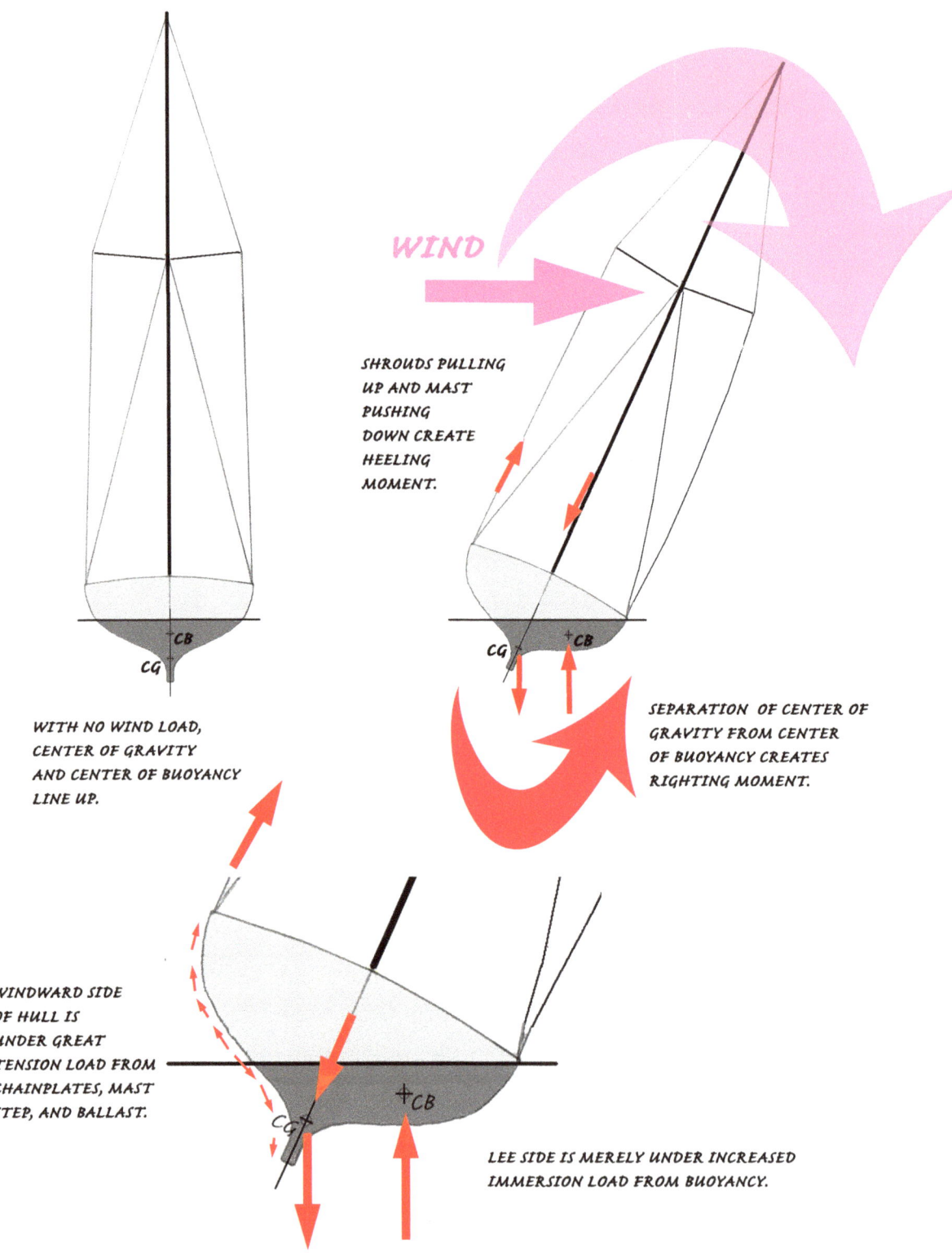

4-7 Wind loads athwartships. Wind on the sails sets up a large heeling moment. When the boat heels, the ballast swings to windward and sets up an equal and opposite righting moment. The moments create a large tension that stresses the windward side of the hull running from the chainplates to the mast step.

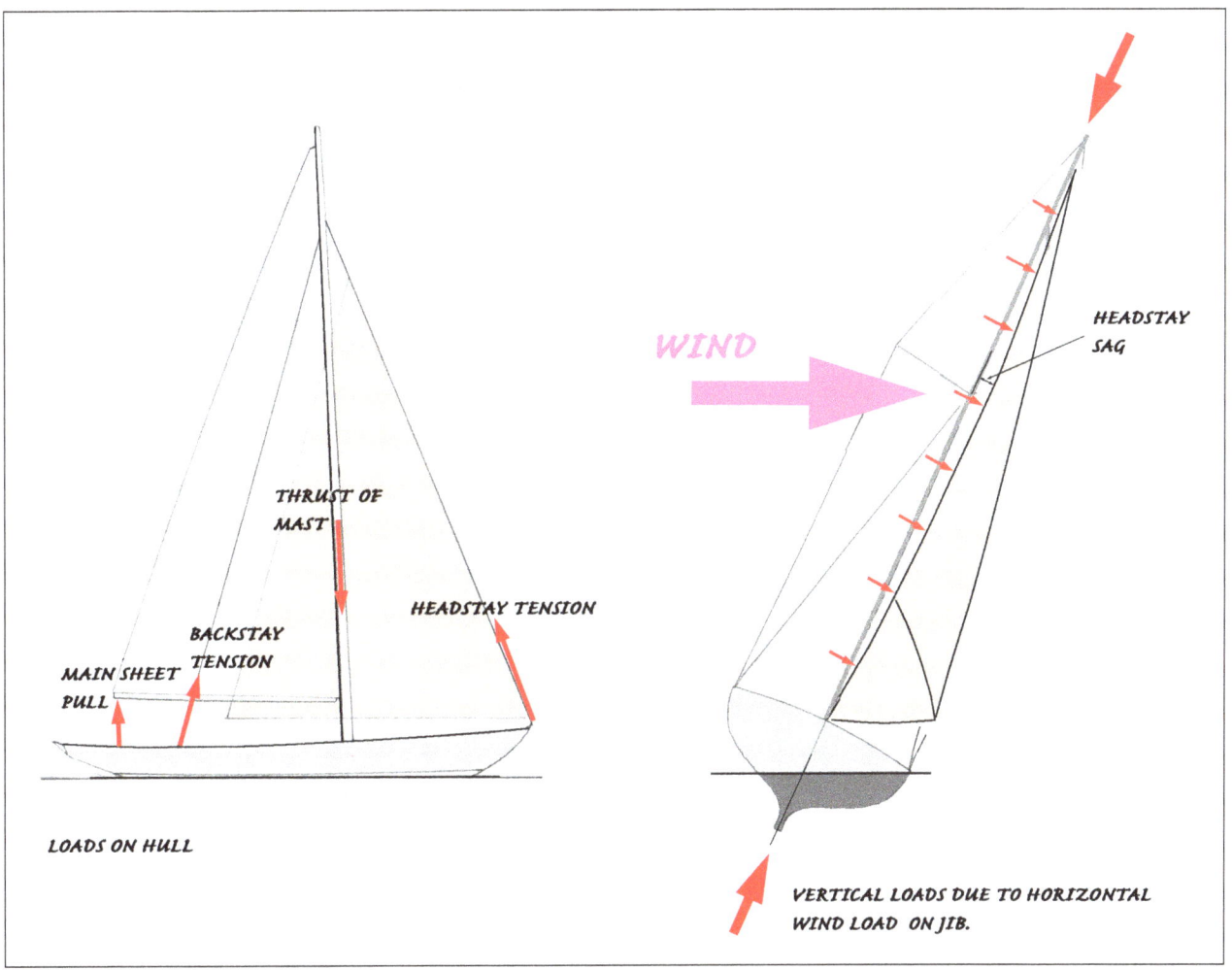

4-8 **Wind loads fore & aft.** Wind acting on the headsail creates tension in the headstay. The headstay tension pulls the mast down and the bow up. It also pulls the mast forward, which is reacted to by the main sheet and the backstay, which pull the stern up as well.

depend on the cross-grain strength of the wood, which, as we know, is low. The coaming is further weakened by a number of large holes for port lights, so this construction is vulnerable in knockdowns. Notice that the wave hits the boat on her windward side, but the damage is caused by the impact with the water on the lee side.

The yawl *OTAIS*, mentioned in chapter 1, on her way to Bermuda one fall was knocked down and split her leeward coaming. Fortunately, her crew got her home. In a similar situation, in the Drake Passage in 1974, the steel 56 foot *SEA LION* was knocked down. Being of steel construction, her house sides had vertical strength, and she

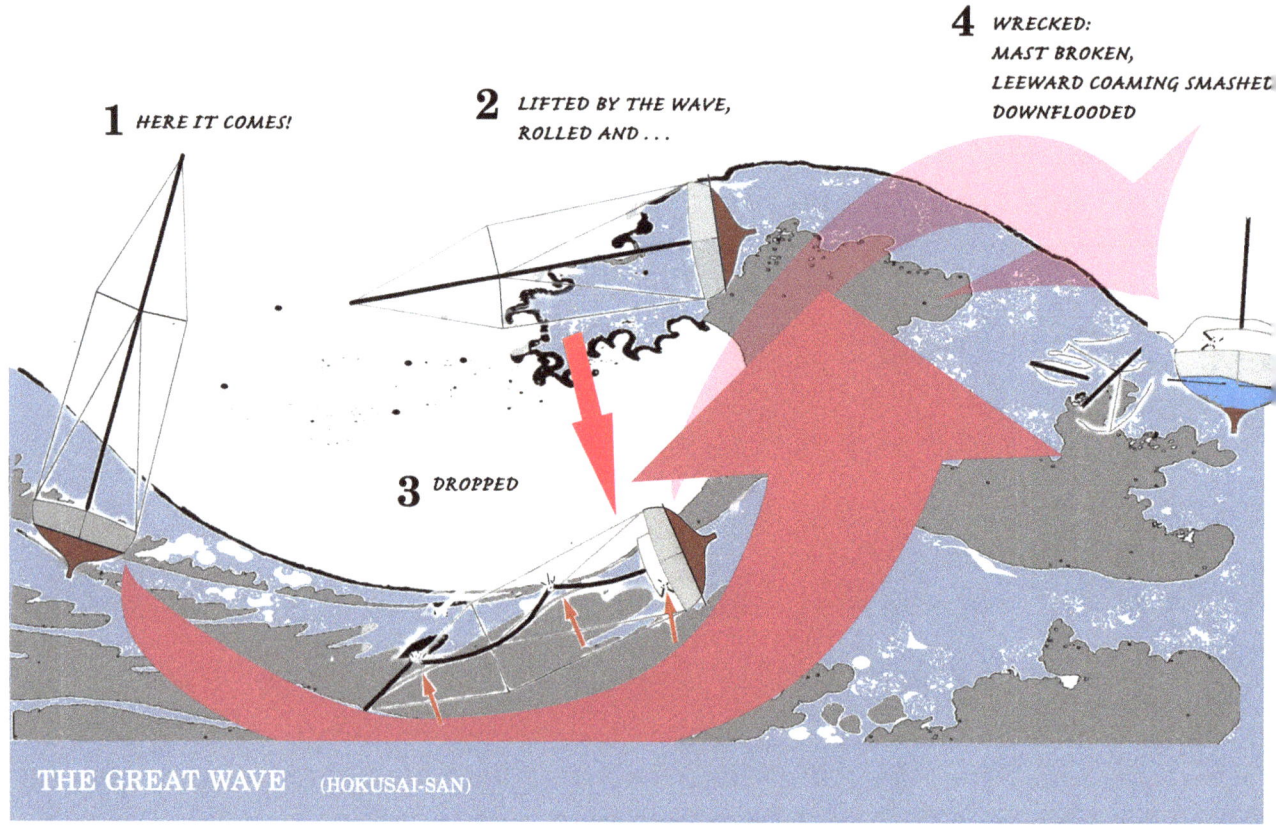

4-9 The knockdown. A breaking wave may lift a boat up and drop it into the trough below. The impact imposes large bending forces on the mast. The mast is designed to resist compression, not bending, so it is liable to break. There are large impact loads on the hull and deck that often split the house coamings of a traditionally built sailboat.

survived without losing her water tightness. Neither boat lost her mast, probably because neither turned all the way under.

Indirect stresses imposed by the sea

There are some indirect stresses imposed on the hull implicit in the direct forces from the sea. We will discuss two, diagonal stresses and hoop stress.

As we saw in the discussion of beams, the bending loads on the hull imposed by gravity, wind, and impact all impose diagonal shear stresses. These stresses are inherent in the nature of bending. Traditional plank-on-frame construction being orthogonal in structure is susceptible to wracking because shearing loads distort rectangles into parallelograms. Both the curvature of the hull and friction between the planks resist wracking, but Herreshoff and other 20th century builders of light craft found it useful to add diagonal metal strapping to resist the wracking forces and so to hold their hulls in shape.

A cold-molded hull built with diagonal laminations has large shear resistance incorporated directly into the shell. There is no need for extra reinforcement and it resists shear deformation with ease.

"Hoop" stress

Hoop stress is a term I have coined to described tensile stresses that run around the hull transversely, stresses that in a barrel would be carried by the barrel's hoops. The greatest of these is the stress applied to the rig by athwartship wind loads, discussed a few pages back. Secondary hoop stresses occur as the deck tries to work loose from the hull and when point rigging loads like jib sheets and such pull on the skin.

The magnitude of hoop stress from the wind loads can be calculated and, in total, is roughly equal to the displacement of the vessel. If we assume the rigging load around the mast is spread over a length of about 1/20 of the waterline (an arbitrary but conservative estimate), we get 20*D/WL pounds per inch as an estimate of hoop strength required. Along the rest of the hull, the loading is less. This strength requirement can be confirmed by comparison with the scantlings called for by Herreshoff's rules. The frames carry the hoop stress through the hull skin of traditional construction. If you reduce the frame's cross-sectional area by a half, to account for weakening due to fastenings and extra loading due to plank swelling, and multiply by its tensile strength, you get a number representing the load each frame can carry around the hull and it is roughly equal to half of the displacement. This number is consistent with the aforementioned 20*D/WL.

The required hoop strength is easily supplied by our cold-molded method because of its diagonal laminated structure. But what happens at sharp corners? The wood laminations will bend around curves but not across corners, such as occur at the deck edge, the keelson, the stem, and, indeed, anywhere there is a chine. This is particularly true of deck structure corners, which have a bad record in knockdowns.

To provide hoop strength around sharp corners, I introduce an additional process. Glass tape is applied, within the wood laminate, around all sharp corners (see figure 10-15.) The corners can be rounded to about ½ inch radius, which the glass tape will easily follow. After the glass tape is applied. the corner can then be sharpened with subsequent layers of the wood laminate. So constructed, the hull will have hoop strength right around its entire section.

Shell thickness

Shells fail when they deform locally and flatten. Then, being like slender columns, they buckle. A boat's hull is a shell structure and we wish to know how thick it must be to be thick enough. There are a variety of opinions, and a variety of

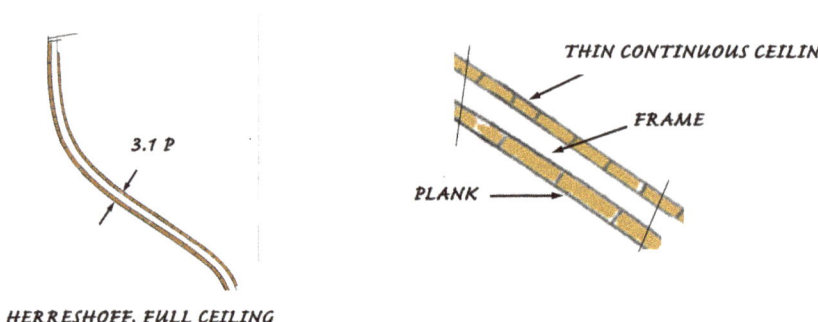

A. HERRESHOFF, FULL CEILING — THICKNESS RATIO ± 115

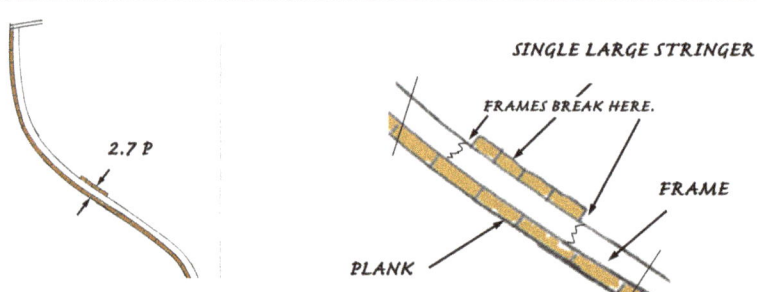

B. NEVINS, ABEKING, OTHERS, SINGLE BIG BILGE STRINGER — THICKNESS RATIO ± 130

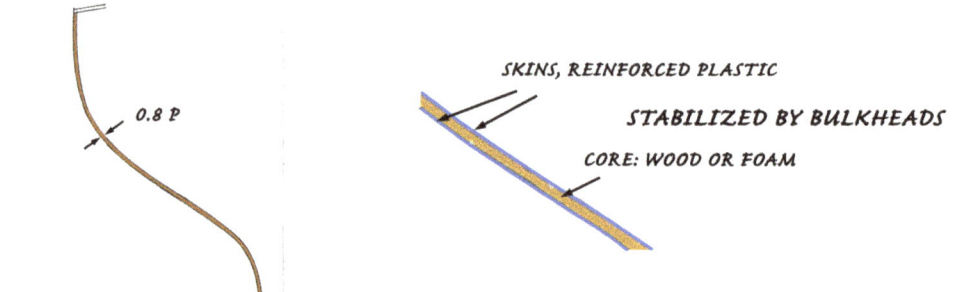

C. CERTAIN CONTEMPORARY BUILDERS, COLD-MOLDED OR PLASTIC — THICKNESS RATIO ± 450

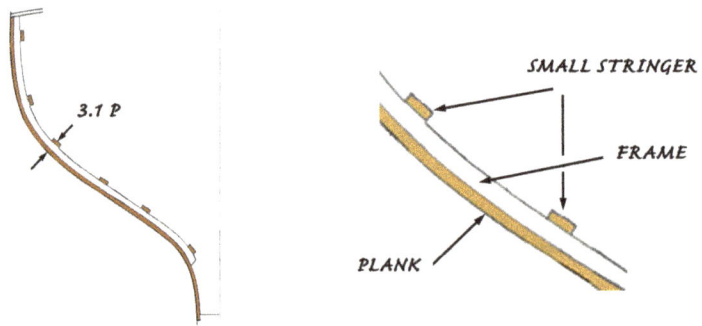

D. PROPOSED STRINGER SYSTEM, SEVERAL SMALL STRINGERS — THICKNESS RATIO ± 115

NOTE: P = PLANK THICKNESS UNDER HERRESHOFF'S RULES

HULL THICKNESS

4-10 Stiffening the hull shell. A comparison of the effective thickness of several hull constructions. Stringers and Herreshoff's structural ceiling act to make the hull into a sort of vierendeel truss that is much thicker than the planking skin.

CHAPTER 4 ◆ SAILBOAT STRUCTURE

plating and framing systems that answer the question in a variety of ways.

Herreshoff, whose boats are noted for their light weight and longevity, used a structural ceiling. His structural ceiling consisted of a second layer of planking fastened to the inside of the frames running unbroken the length of the hull. His construction where the two wood skins, plank and ceiling, are spaced apart by the frames is about three times the thickness of an unceiled hull (see figure 4-10 A).

Herreshoff's structural ceiling was not widely imitated. Rather, most plank-on-frame yachts used a single, large, bilge stringer running along the inner side of the framing about midway between the sheer and the garboard. By concentrating Herreshoff's inner skin into a single piece, it is easier to build and most of the inner surface of the hull remains exposed to sight—a good thing (**RULE 3**). A drawback is that the heavy stringer concentrates bending stresses on the frames at its edges, frequently breaking them—not such a good thing (see figure 4-10 B).

Herreshoff's structural ceiling gives a hull shell with a radius/thickness ratio of about 115:1. The single bilge stringer system gives a ratio of about 130:1. Typical metal boats are about the same. But many contemporary boats built with sandwich construction are much thinner, having ratios of as little as 450:1 (see figure 4-10 C). Perhaps their structural bulkheads make the difference. Whether such thin shells are wise or not, time will tell.

Cold-molding changes the fundamental nature of wood boat construction by amalgamating the frames and planks into the laminate. Cold-molded hulls readily carry tensile loads and hoop stresses. The main structural question with a cold-molded hull, is then, how thick need the hull shell be to maintain its geometrical integrity? There is little literature on the subject of hull plating thickness other than the scantling rules. I think the question is open.

STARRY NIGHT has a stringer-generated hull, of which we will talk more later. The stringer-generated hull replaces Herreshoff's continuous ceiling with a series of smaller stringers spaced along the frame. They neither conceal the inner surface nor concentrate large loads on the frames. Used with cold-molding, the frames stiffen the shell athwartships but, most importantly, space the stringers out from the hull shell, increasing its thickness to that of a Herreshoff hull. It does so with little added weight and several construction advantages (see figure 4-10 D). I offer it as an interesting possibility.

We will return to stringers in chapters 6 and 7. But first, we have continued to talk about benefits of wood construction, so let's look at it in detail and see why we wish to use wood for our oceangoing sailboat.

CHAPTER **5**

Why Wood?

EVERYONE LOVES WOOD. What wonderful stuff it is. With its beautiful grain patterns sometimes straight, sometimes swirling, each species different, each piece different, in color, weight, and texture. But not helter-skelter different, rather well-ordered different, telling the tale of its growth in the forest, its conversion into lumber, and its final shaping by the carpenter. It takes a hard man to be unmoved by the medullary rays of a fine piece of rift-sawn white oak, or the smell of camphor, cedar, and resin, or the joinery of a paneled door that holds parts of changing width into a panel of constant dimension.

But to build a boat with it? How can it even be done? The sailor must put away his sentimental love of beauty. The sea is demanding. He must be practical. And so it is. It turns out that, when you look at it closely, wood is the finest material of them all for boatbuilding in a number of nice ways. Since the beginning, boats have been made of wood. The reasons are that wood is strong, plentiful, and easy to shape. It has the right surface qualities. As we will discover in chapter 6, it has the correct density. In truth, it makes the best boat.

CHAPTER 5 — WHY WOOD?

The strength of wood as a structural material is not generally understood. At first glance, wood seems sort of weak. You can break a stick across your knee. But the fact is that, when its weight is taken into account, wood is amongst the strongest of common structural materials. It is exceeded, today, only by exotic carbon fiber composites and even then only under special conditions. Wood's strength is not so surprising when one considers that wood evolved as the structural material of trees, the largest of all living things. More on wood's strength in the next chapter.

For all history, wood has been the domestic material of choice (the *Three Little Pigs* notwithstanding!) The engineer may discount this, but the designer must not. The oceangoing sailboat is home at sea for her crew. They work her deck and rest in her interior. They are in constant contact with her surfaces and the nature of these surfaces matter. The crew holds on to them, sits on them, lies down on them, sleeps amongst them, walks on them, climbs them. With wood surfaces, they do this happily. Metal and plastic surfaces do not provide the crew such comfort. They are cold, wet, noisy, and slippery. Perhaps, for professional racers, cold, slick, wet metal and plastic surfaces are adequate—sailing such boats is their job and they are paid to do it. But for the man who sails for pleasure, boats made of metal and plastic need an additional, wooden boat built inside them to make them habitable (**RULE 2**).

Wood is a thermal insulator, poor at absorbing or radiating heat. Under the tropic sun it does not burn the skin, nor will it suck heat from hands on a cold night. Wood stays close to the temperature of the air around it, so a wood hull and deck resist condensation in the cold and damp. A small stove warms the wood cabin quickly and cheerily. Wood sounds good; it muffles sharp noises and does not ring.

A wooden boat has a habitable interior without the need to cover the structure with insulation or décor, which is more than an aesthetic concern. Direct exposure of the structure to view not only makes a beautiful decor, it is a requirement of the sea (**RULE 3**). Being exposed, the structure is under constant surveillance. Damage, especially slowly developing damage, is immediately apparent, and no barrier stands in the way of its repair. The builder of metal or plastic hulls must ceil them with an insulated liner to make them comfortable and pretty. More than just adding weight, the ceiling hides deterioration and renders it inaccessible for repair—an unsafe condition.

Wood boats are easy to repair. They can be built with hand tools and they can be repaired with hand tools, tools that can be carried aboard. Material to make repairs can be obtained in the far corners of the earth. Wood is good for the self-sufficient sailor's boat (**RULE 6**).

One last point, when we build a boat of wood, we are not limiting ourselves to a single material. Wood is a family of materials. Each species of wood has particular physical characteristics that can be selected to fit the specific demands of each element of the vessel. The variety increases the performance of the whole and adds to the beauty of the craft.

On the *Alerion Class Sloop* we used 8 different species, each selected for its special properties to do its particular job in the best possible way.

Cedar, eastern white, is lightweight, durable; used for inside planking layer.

Fir, Douglas, is strong in bending and tension for its moderate weight; used for core planking, deck plating, and booms.

Mahogany, honduras, is medium weight, hard, beautiful; used for exterior planking, coamings, interior accommodation.

Spruce, sitka, has highest stiffness for its weight; used for the mast.

Teak is beautiful, durable, can be exposed to weather without finish—it has a non-skid surface, making it particularly useful for soles and cockpit seats.

Pine, white, is light, strong and inexpensive; used for bunk bottoms and shelving.

Ash, white, is heavy and very strong; used for structural framing, stem, tiller.

Lignum vitae is very hard and slippery; used for abrasion pads at jib sheet cleats. Unfortunately it is, today, unobtainable in the USA.

Why not wood?

All these advantages of wood for boat construction raise the question, why did anyone bother to develop other materials for boat construction. The answer lies in two particularities of wood's structure and one problem with traditional technology.

The first particularity of wood is that it is an organic material and highly biodegradable. The sea being the world's stomach, biodegradability is a serious vulnerability. Rotten boats are not strong (**RULE 1**).

The second peculiarity of wood is that it is a linear material with highly directional properties. Coming from trees, it has a grain structure that runs parallel to the trunk or branch. Wood is about 20 times as strong parallel to its grain than it is across the grain. Wood is so dimensionally stable parallel to the grain that it is used for clock pendulums, but it changes size as much as 10% perpendicular to the grain as its moisture content changes. All wood structures must accommodate wood's anisotropic qualities; boats are no exception.

So how do you assemble this linear material to form the shells required for boat hulls? A shell demands strength in both directions, longitudinal and transverse; wood offers it in one. Over the centuries, the maritime nations of the West settled on the technique of building the hull shell by crossing longitudinal planks with frames and holding the basket together

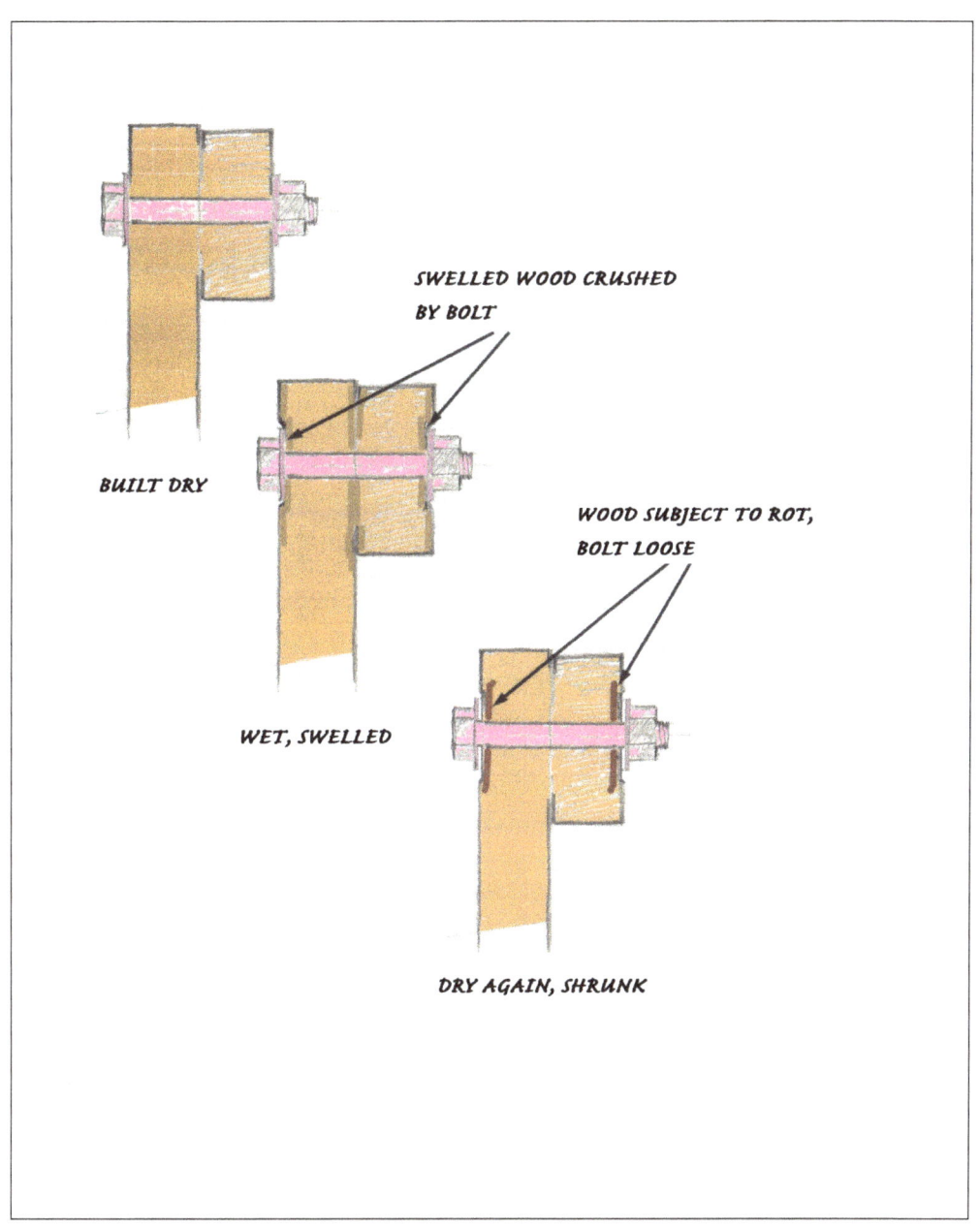

5-1 **Bolted connections.** Bolted connections work by squeezing two surfaces together so that friction prevents them from moving. But the wetting/drying cycle shrinks the wood and leaves the bolts loose. It also damages the wood surrounding the bolts, making it susceptible to rot.

by metal fastenings. This technique works but has a grievous fault. Over time, especially in a saltwater environment, the metal-fastened joints fail (**RULE 4**).

Metal and wood are antagonistic because they have very different physical properties. Metal conducts heat and wood does not. Metal is hard and wood is soft. Metal is much stronger than wood is across the grain. Furthermore, metal is stable under changes of humidity while wood is not. Wood parts swell with increasing moisture content; metal fastenings do not. Over time, the relative movement between the two materials degrades the joint.

The process is that new wood is bolted (or screwed or riveted) together into a tight sound joint. All is well till the wood gets wet and expands. The metal fastener does not expand and, being stronger than the wood, it squeezes the wood beyond its elastic limit (see chapter 6), crushing it. Having been squeezed beyond its elastic limit, the wood does not fully rebound when it dries; it shrinks to less than its original thickness, leaving the joint loose.

A fact not widely understood is that bolted joints function properly by loading the bolt not in shear, but in tension, so that it squeezes the two joined members together. The bolt need not even touch the sides of its hole; its purpose is to squeeze. Squeezed together, friction binds the joined pieces and prevents them from moving (rotating or slipping) relative to each other. When expansion and contraction of the wood loosens the joint, the friction disappears leaving the joined parts free to rotate and slip. The wood at the edge of the fastener holes is crushed and worn away by the metal. The joint has failed.

N. G. Herreshoff preferred to use screw fastenings rather than bolts or rivets in part because he believed that screws, not going all the way through the frame, are less affected by the shrinkage and expansion of wood. There may be some truth to this. Screws, however, suffer another process. As the wood expands and contracts, the sharp threads of screws cut the wood they grip, eventually boring out a hole slightly larger than the screw itself and leaving it without bite. Vibration induces the same effect as expansion/contraction. Anyone who has used a modern vibrating "Multitool" can visualize a little bit of saw dust falling off the sharp screw threads each time a wave hits the hull. One reason for Herreshoff's hulls' reputed longevity may be that his hulls are much stiffened by their structural ceiling, so they do not vibrate as much; hence the "Multitool" action of his fastenings is diminished. I don't know, but it is an interesting idea.

Other drawbacks of metal fastenings include the fact that they conduct heat much better than the wood parts they join. When they get cold, water vapor condenses around them. Drawn deep into the interior of the wood parts, the condensation promotes rot. Conversely, wood, particularly the white oak commonly used for

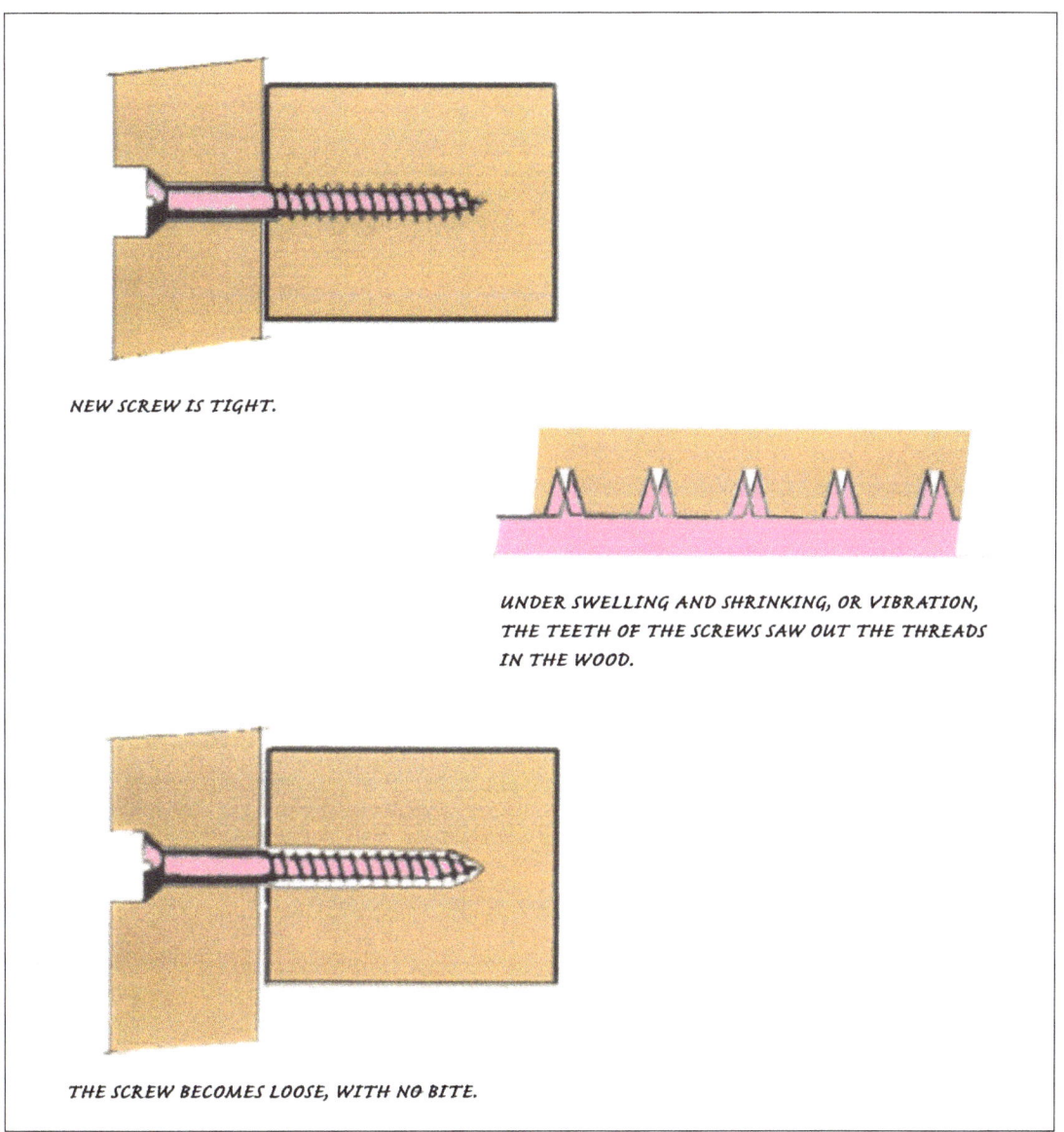

5-2 **Screw connections**. Screws lessen the crushing that bolts cause, but the wetting/drying cycle and vibration work to destroy the integrity of the connection in another way.

frames, contains acids that corrode metal fastenings. Although bronze and copper will resist this, iron fastenings will not and will rust to nothing within a decade or two.

Lastly, metal fasteners require substantial holes, which significantly weaken the pieces they fasten. In the bilge around the mast step there often is insufficient space to make the parts big enough to both carry the load and absorb the fastening holes.

Enter cold-molding

We all like wood boats, but we don't like that they wrack, leak, rot, and fall apart, nor that they fall apart more quickly when driven hard or left unattended. While proper design and use of the finest materials can delay the demise of the traditionally built wooden boat, they cannot eliminate the difficulties of using a linear material to form a bidirectional structure, nor will they eliminate the antagonism between the wood structure and the metal fastenings holding it together. But cold-molding and epoxy bonding will bypass these faults in traditional construction.

Which is why cold-molded construction caught our attention when my brother and I were introduced to it in the 1970's. We realized cold-molding and epoxy bonding could eliminate all the disadvantages of wood construction and enable construction of the finest boats imaginable. Cold-molding eliminates metal fastenings, the bugaboo of traditional plank-on-frame construction, and offers an easy way to construct a bidirectional shell from unidirectional wood. The technique allows the sailor to have his cake and eat it too.

Cold-molding begins with plywood. Samuel Bentham, a British naval engineer, patented the idea of plywood in 1797. Fifty years later the veneer lathe was developed to peel veneers for plywood off a log like unrolling a roll of paper. Early forms of plywood were on the market in the late 19th century. After the veneer lathe was invented, the chief technical problem with plywood was the glue. The first marine glues appeared on the scene late and required both heat and pressure to cure. Heat and pressure were fairly easy to apply to flat sheets of modest size that were stacked up, pressed, and cured in ovens, presenting the world with marine plywood. Molded marine plywood was considerably more difficult. The ovens required for curing large, curved parts, so-called "molded plywood," which was developed by the aircraft industry during WWII, were expensive. So was the tooling to hold the veneers in place while they cooked under pressure. But for large production runs of small boats, it worked, producing a very good hull. After the war, Uffa Fox's sailing dinghies were made of this [hot-]molded plywood.

In the 1970's, epoxy resins became available to boatbuilders, changing the game. Epoxy glue is strong, waterproof, and gap filling. Further, without pressure, it cures at room temperature.

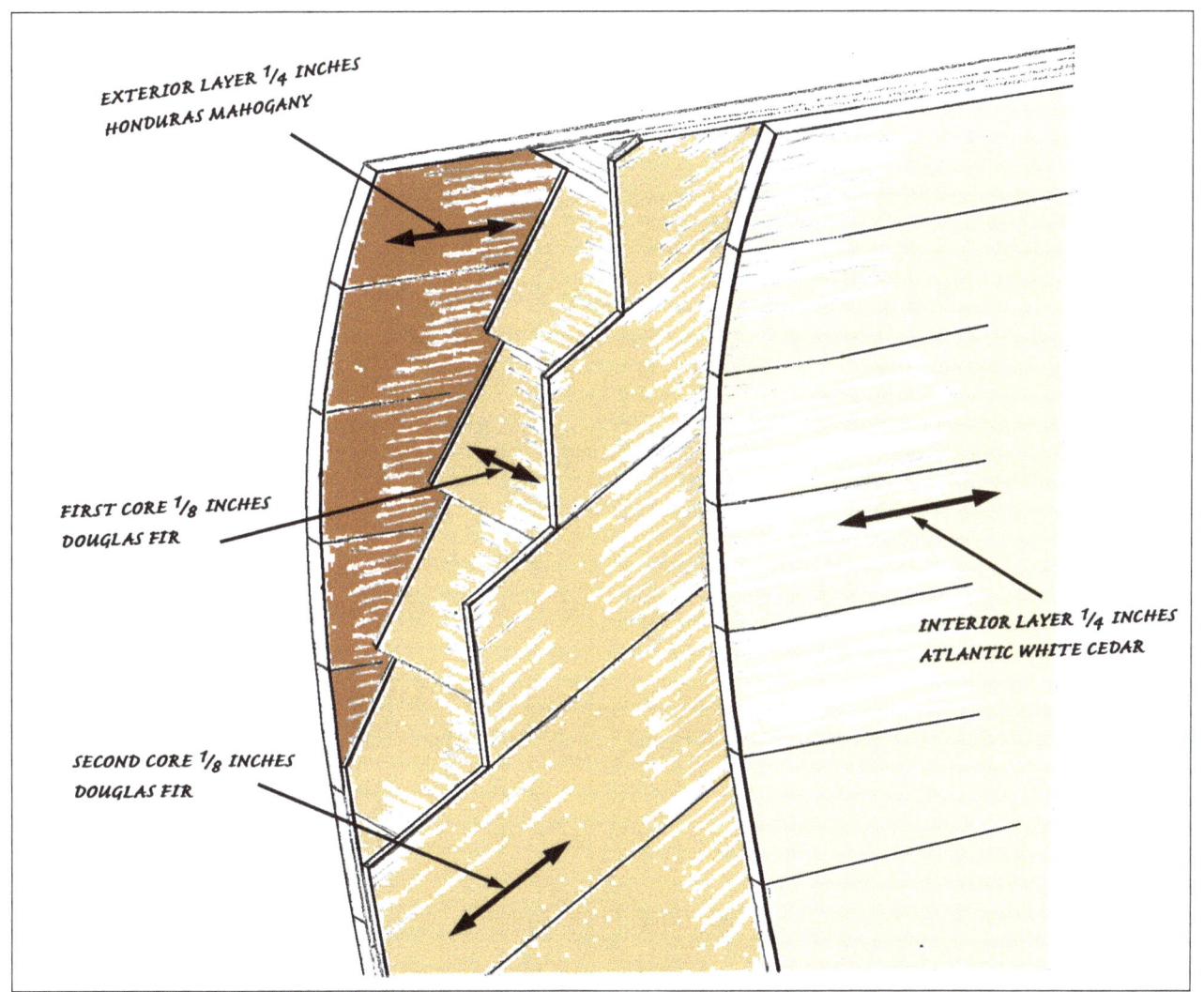

5-3 **Cold-molding.** This cutaway drawing of the *Alerion Class Sloop* laminate shows the four layers of cross planking that make a wooden skin strong in both directions. The work of traditional frames is done by the core diagonal layers and epoxy glue takes the place of metal fastenings.

It degrades only by sunlight. With its introduction, cold-molded plywood (shortened to "cold-molded") became suitable for the small boat shop. Ovens are not necessary. Tooling is light and easy. The epoxy glue is compatible—chemically, mechanically, and thermally—with the wood it bonds.

Just how does cold-molding replace the traditional plank-and-frame hull shell? Well, first look at how the traditional hull shell works. The planking carries the fore & aft loads and it does so with great strength. The planks also carry transverse compression loads, although with much less ability as their cross-grain strength is only about $\frac{1}{20}$ of their longitudinal strength. Further, the planks are unable to carry any transverse tension load since they are interrupted by joints every few inches. Transverse tension loads fall entirely on the frames, which must carry all the hoop stresses. To remind, those include loads due to swelling and caulking of the planks, ballast and rigging loads, and random impact loads caused by waves.

Differing from traditional construction, cold-molding consists of multiple layers of planking, running in cross directions, glued together into a laminate. There are lots of choices, but generally the laminate consists of fore & aft layers carrying the fore & aft loads crossbanded by diagonal layers that carry the hoop stresses. Importantly, there are no fastenings and the whole is bonded into a single unitary piece with epoxy resin, which appears, after my sixty years of experience, to have an unlimited lifetime.

There are a number of different laminate designs. That for the *Alerion Class Sloop* consisted of an inner and outer layer of ¼ inch fore & aft planking with two interior layers of ⅛ inch running diagonally. This resulted in about 80% of the material running longitudinally and a carvel-planked exterior surface suitable for bright finish. Many boats today consist of a thick fore & aft layer in the inside covered with several layers of thinner diagonal planking outside. These boats are usually painted.

The importance of cold-molding and epoxy bonding is that they eliminate the drawbacks of wood construction. By doing so, they open the way for something new and delightful, a wooden oceangoing sailboat that meets all the rules—a beautiful, seaworthy home for the sailor at sea.

CHAPTER 6

Wood as a Structural Material

TO TALK ABOUT WOOD as a structural material, we are going to have to go into the concept of strength in a bit of detail. We have talked in chapter 4 about strength from a geometric point of view, and now, we will talk about it from a material point of view.

Like most boatbuilding materials, wood is an elastic material with some ductility in compression, but it differs from the metals and plastics in its directionality. Wood is about 20 times stronger and stiffer parallel to the grain than it is across the grain. Dimensionally it is very stable along the grain and quite unstable across the grain. With changes of moisture content, its width can change as much as 10%. This directionality is not always a disadvantage. Wood's weakness cross-grain is offset by its immense strength with the grain; we just need to be aware of it and take advantage of it.

Strength

Strength is a general term that refers to a number of different properties. It includes **tensile** strength (resistance to being pulled apart), **compressive** strength (resistance to being crushed), **flexural** strength (resistance to bending), **stiffness** (resistance to deformation), **hardness** (resistance to impact and puncture), **fatigue** (resistance to weakening when bent

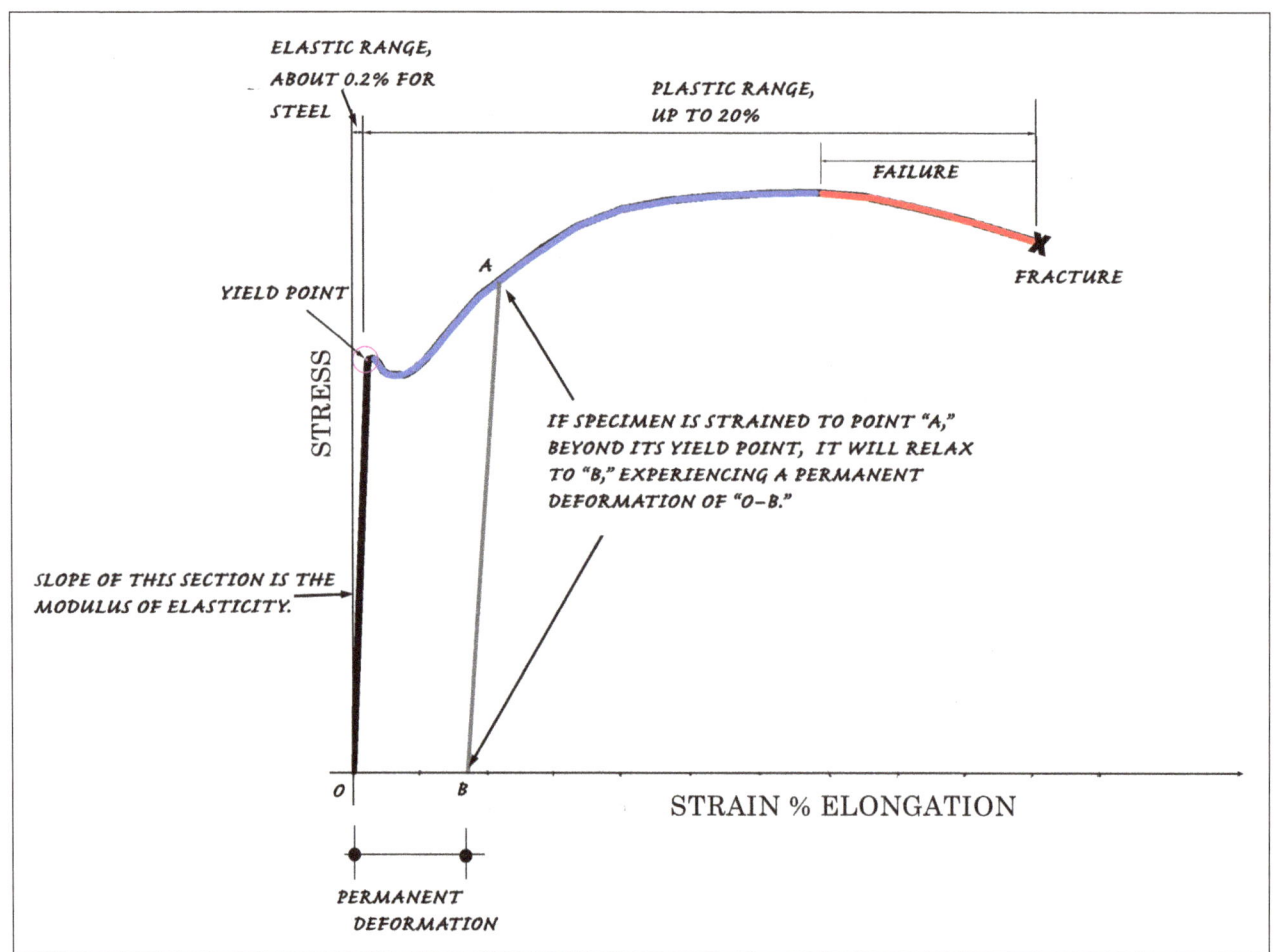

6-1 Stress-strain curve. This curve is for mild steel, representative of ductile materials such as metals and plastics. The stiffness of the material, measured by the constant called its modulus of elasticity, is shown by the slope of the black, straight section of the curve.

When the yield point is reached, at about 0.2% elongation for steel, the sample does not break, but stretches about 10 to 20 times farther, carrying high load before finally fracturing. This is the plastic range shown in blue and then red. Non-ductile materials like glass or wood in tension break at the yield point.

back and forth), and **durability** (resistance to deterioration in service). All of these strengths are desirable in material used to build a sailboat. The two most important ones are bending strength and stiffness.

Before going further, to get the language straight, here are three definitions:

load   a push (compression), a pull (tension), measured in pounds of force

stress   a load applied across a sectional area, measured in pounds per square inch

strain   deformation of an element in the direction of a load, measured in inches per inch or the percentage change in length

Stress and strain are related.[1] The more you stress, pull, or push a specimen, the more you strain it. The relationship may be graphed by plotting strain against stress. The shape of this stress-strain plot tells much about the material being tested. The stress-strain curve for elastic, ductile materials (metals, plastics, and fiberglass) is similar to that shown in figure 6-1 which is the diagram for mild steel.

Characteristic of the curve is the straight steep rise to the yield point at the beginning. This is the "elastic" range where the material acts like a spring. The elastic range has two properties:

1. Strain is proportional to stress.
2. When the stress is released, the specimen returns to its original size.

In the elastic range, the black line, the slope of the straight steep part of the curve, is a constant, the **modulus of elasticity** of the specimen. The modulus of elasticity, symbolized "E," measures the stiffness of the material.

As the stress on the specimen is increased beyond the **yield** point, two new qualities express themselves:

1. The strain ceases to be proportional to stress. A slightly increasing load stretches the specimen a long way. The specimen is now in the "plastic" range, which may be 100 times the length of the elastic range.
2. When the stress is released, the specimen does not return to original size; rather, it is permanently deformed.

The stress-strain curve is found with a testing machine. A testing machine grips a sample, in this case a steel bar, and pulls on it with a hydraulic ram. As it pulls, one gauge measures the pull on the bar, its stress, and another measures how far the bar stretches, its strain. The stress-strain curve is a graph of the load versus

---

1. A relationship discovered by Robert Hooke in 1660 and eponymously named Hooke's Law. It is not really a law; rather, it is a description of Hooke's observations. But over the years it has inverted into a definition of the word "elastic" as used in "elastic range," "elastic limit," and modulus of "elasticity."

the stretch. The steel bar being strong, it requires a lot of load to stretch it very far. You cannot really see the strain with the naked eye.

Think of the steel bar as a spring. You can pull on a spring and stretch it; the harder you pull it, the farther it stretches. When you relax the pull, it reverts to its original size. But if you pull it too hard, it stretches beyond its elastic limit, and some of its coils will unwind. The spring is in its plastic (blue) range. It will stretch a long way and the unwound coils will stay unwound when you quit pulling. The spring will retract, but not all the way back to its original size. The unwound coils are permanently deformed. Finally, if the spring is pulled even farther, it straightens out entirely and ultimately breaks (red line).

The stress-strain diagram for wood in compression looks similar to steel. In tension, however, wood has no plastic range. At the yield point, it breaks.

The stress-strain curve for common boatbuilding materials has been determined in the lab by testing machines. The tests give us the two constants, yield stress and modulus of elasticity, which are indexes of bending strength and stiffness respectively. We will use these constants to compare the different boatbuilding materials as we look to see which will do the best job for us.

Table 6-1, columns 3 and 4, shows yield stress and modulus of elasticity for a variety of boatbuilding materials including different species of wood.

Specific strength

Since weight restricts all things that move,[2] including sailboats, it must be taken into account when choosing a material. Usually a structure of high strength for its weight will perform better. So when choosing a material for strength, it is important to know if it is strong for its weight. If we adjust the strength constants by dividing them by their material's specific gravity, we get specific strengths, a measure of the strength of the material adjusted for its weight. Columns 5 and 6 of the table show the constants after such adjustment. Somewhat surprising to our intuition, wood scores high, and certain species take the prize.

Interestingly, the stronger woods, steel, and aluminum are all quite close together, in both specific bending strength and specific stiffness. The outliers are fiberglass, which is quite inferior in stiffness, and carbon fiber, which is almost twice as strong, pound for pound, as the best of the other materials.

The peculiar effect of density

We learned in chapter 4 that the actual strength of a structure is dependent on both its geometry and its material. And Timoshenko's work showed us the importance of thickness for structural efficiency. Strength increases faster than

---

2. Uffa Fox famously said, "The only place for weight is in a steamroller."

thickness. Thickness increases the strength-to-weight ratio of a structural element. This is true for beams, shells, and columns, and it leads to a peculiar effect of density.

When I was a student at MIT in the 1960's, I was taught structures by William LeMessieur. LeMessieur engaged us with his "laws" of structures. His laws brought deep structural principles to the surface with simple rules of thumb. Professor LeMessieur was a fine fellow and a superb teacher. His laws were general and had a hard basis in science, so they held true and were useful over a wide range of conditions.

Well, I am no MIT professor, but I do have a rule about density that is a bit more intuitional than LeMessieur's, but which I think has some merit. I offer that:

With materials of similar specific strengths, the optimum density for the material of a structural element is correlated to the load density that the element is subject to.[3]

This is so because, for equal weight, elements of lower density may have greater thickness and hence greater structural efficiency. On the other hand, higher load density implies a lack of space. The lack of space leaves no room for additional thickness, so the load must be carried by a stronger material. For materials of similar specific strength, this means a denser material.

A practical example of needing higher density in highly stressed areas is the migration from spruce to aluminum for mast construction. Spruce and aluminium have the same specific strength, but aluminum is about 6 times denser. In the days of low aspect ratio sail plans, load densities on spars were lower and wood made practical masts. With the advent of the Marconi rig, rigging loads became much greater, generating high load densities where the rigging met the spar. With spruce spars the attachment is accomplished with elaborate sheet metal tangs diffusing the load over a large area. The denser aluminum does the same job with simple bolts, greatly relieving the crowding at the masthead and hounds, where loaded shrouds, stays and halyards all come together within a limited space. The denser aluminum, requiring less space for structure, leaves more room to make the connections than bulkier spruce. High load concentration calls for a denser material.

Over the years boatbuilders have learned that the action of the sea and wind on moderate-size boats asks for a hull material density of about 0.5. Interestingly, while boats are best built of materials with density of about 0.5 (wood), aircraft seem to be best at 2.0–3.0 (carbon and aluminum), and land-based vehicles, bicycles, cars, and trains are at 6–8 (steel).

---

3. A note of caution: "load density" is my phrase and is not a formal engineering term. I offer it not with a strict definition but as a useful concept. Think of it as the **load** relative to the amount of **space** available to accommodate it.

TABLE 6-1 Mechanical constants for boatbuilding materials.

	Density	Flexural Strength, psi	Modulus of Elasticity, psi	Specific Flexural Strength, psi	Specific Modulus of Elasticity, psi
Woods					
Atlantic white cedar	0.32	6,800	930,000	21,250	2,906,250
Butternut	0.38	8,100	1,180,000	21,316	3,105,263
Red spruce	0.40	10,800	1,610,000	27,000	4,025,000
Douglas fir	0.48	12,400	1,950,000	25,833	4,062,500
Honduras mahoghany	0.54	12,100	1,280,000	22,407	2,370,370
White ash	0.60	15,000	1,740,000	25,000	2,900,000
Teak	0.66	15,400	1,450,000	23,333	2,196,970
White oak	0.68	15,200	1,780,000	22,353	2,617,647
Ipe	0.76	16,200	1,650,000	21,316	2,171,053
Greenheart	1.03	26,200	3,040,000	25,437	2,951,456
Metals					
Aluminium, alloy 5456	2.65	53,000	10,000,000	20,000	3,773,585
Steel, alloy 1040	7.83	89,000	30,000,000	11,367	3,831,418
Stainless steel, alloy 316	8.03	100,000	28,000,000	12,453	3,486,924
90-10 coppernickel, alloy 706	8.94	50,000	18,000,000	5,593	2,013,423
Silicon bronze, alloy CDA 655	8.36	63,000	17,000,000	7,536	2,033,493
Composites					
Glass/polyester, 25 oz. woven roving, 10 oz. cloth	1.69	35,000	1,960,000	20,710	1,159,763
Carbon/epoxy, unidirectional	1.55	300,000	15,000,000	193,548	9,677,419
Carbon/epoxy, sheet laminate	1.60	100,000	10,150,000	62,500	6,343,750

Items tinted in red are superlative.

Simple plates

By looking at flat plates of constant weight, made of different material, we can see the effect of their density on their strength. It is, in fact, quite amazing.

A 1 inch fir plank weighs the same per square foot as a ¹⁄₁₆ inch thick sheet of steel, ³⁄₁₆ inch of aluminium, or ⅜ inch of fiberglass. As a simple plate, the wood plank has 30 times the breaking strength and 4,000 times the stiffness of the steel sheet.

Table 6-2 shows this numerically for the other materials, giving the relative bending strength and stiffness of each. For illustrative

Units	Authority
lbs/in^2	*Encyclopedia of Wood*
lbs/in^2	*Encyclopedia of Wood*
lbs/in^2	*Encyclopedia of Wood*
lbs/in^2	*Encyclopedia of Wood*
lbs/in^2	*Handbook of Hardwoods*
lbs/in^2	*Encyclopedia of Wood*
lbs/in^2	*Handbook of Hardwoods*
lbs/in^2	*Encyclopedia of Wood*
lbs/in^2	*Handbook of Hardwoods*
lbs/in^2	*Handbook of Hardwoods*
lbs/in^2	*Handbook of Oceanographic Engineering Materials*
lbs/in^2	*Handbook of Oceanographic Engineering Materials*
lbs/in^2	*Handbook of Oceanographic Engineering Materials*
lbs/in^2	*Handbook of Oceanographic Engineering Materials*
lbs/in^2	*Handbook of Oceanographic Engineering Materials*
lbs/in^2	*Marine Design Manual for Fiberglass Reinforced Plastics*
lbs/in^2	Author's estimate
lbs/in^2	Author's estimate

purposes I have chosen plates that weigh 3 pounds per square foot, about the weight of typical hull plating for a 40 foot sailboat.

The plates of lighter materials, being thicker, outperform those of heavier materials dramatically, even those of carbon fiber laminates. The heavier materials must use sandwich or framed structures; else, they are too weak for practical consideration. Effectively, framing or sandwiching makes the structure itself less dense and allows the heavier materials to compete with wood. Wood just happens to be in the sweet spot. The traditionally constructed boat has a shell that is made of a single thickness of planking that just happens to be not too heavy, not too light.

The structure of a cold-molded laminate

The cold-molded laminate is made of layers of wood running in cross directions. Under changes of moisture content the layers swell across the grain but not parallel to the grain. Why doesn't the differential swelling break the laminate apart? Figure 5-3 illustrates the laminate and the double arrows indicate the direction of the grain in the various layers.

I am aware of no scientific study of this question, but I believe what happens is this: as the cross-grain wood gets wetter, it tries to expand, but it is restrained by the neighboring parallel-grain layers. The cross-grain layer will strain the parallel grain layer. But the parallel-grain layer is so much stiffer and stronger than the cross-grain layer, it will bring the cross-grain layer to its yield point long before it, itself, is much strained. The cross-grain layer, as it tries to swell, will yield in compression to conform its size to that of the parallel-grain layer.

The glue must be strong enough that the parallel grain layer will constrain the cross-grain layer. The glue will fail only if the cross-grain

TABLE 6-2 Bending strength of simple plates of materials of different density.

Material	Density	Thickness	Relative Strength	
			Bending	Stiffness
Atlantic white cedar	0.32	2.40	123%	222%
Douglas fir	0.48	1.60	100%	100%
Glass/polyester, 25 oz woven roving, 10 oz cloth	1.69	0.46	23%	2%
Aluminum, 5456	2.65	0.29	14%	0.4%
Steel, 1040	7.83	0.10	3%	0.02%
Carbon/epoxy, bidirectional	1.60	0.48	62%	4%

Items tinted in red are superlative.

layer is so thick that its yield stress times its thickness is greater than the strength of the glue. Since epoxy has a working strength of about 1000 psi and the common planking woods have cross-grain yield of about 400 to 800 pounds per square inch, laminates with layer thickness of less than 1 inch should be stable with some factor of safety in reserve.

Some actual testing research on the question would be helpful.

Wood as a material is strong for building boats. Further, its density allows us to build a boat with a relatively simple structure. Cold-molded laminations allow us to fashion strong and stable hull shells. In the next chapter we will compare traditional wood boat structure with contemporary structures having metal and plastic hulls and compare them all with a new structure, the stringer-generated hull, used with cold-molding.

CHAPTER 7

The Stringer-Built Boat

THIS BOOK BEGAN WITH the story of the early days of ocean cruising in yachts and paid some homage to the sailors and designers who developed the art of building and sailing them. It continued to discuss what the sea and sailor require of a boat to make ocean voyages. To get a better understanding of the issues involved, we digressed to look at the mechanics of structures, the particular loads on seagoing yachts, and the particular advantages of using wood as a material for building those yachts, especially using the still new cold-molded and epoxy-bonded construction technique.

The remainder of this book is going to assume that you, the reader, have come to share my interest in using cold-molded wood and epoxy bonding to make a superlative small ocean-cruising boat. In the next chapters I brazenly introduce a general system for building the blue water sailboat that has evolved in my mind over the past forty years of building and sailing wood boats.

In essence, the hull of a sailboat is an elongated watertight shell. Its principal loadings are fore & aft acting as a beam. The blue water sailboat adds a watertight deck to the shell, forming a tube. The tube form makes a strong beam as long as it has sufficient hoop strength to hold itself together and sufficient surface stiffness to hold its shape and avoid buckling.

Longitudinal framing

There are two obvious ways to stiffen the surface of a tube: one can use frames running around the section of the tube, or one can use longitudinals running along its length. Longitudinals have two significant advantages over frames. Running longitudinally, the material of the longitudinals adds to the strength of the tube as a beam. Also, there are fewer of them and they are easier to make because their shape can be formed by bending rather than sawing.

I imagine that the idea of using longitudinals to generate boat hulls goes way back in time. I first learned about longitudinal framing reading, in Francis Herreshoff's magnificent biography of his father, the description of America's Cup defender CONSTITUTION, built in 1901. She was longitudinally framed, and Francis writes of the technique:

> This is the lightest known system of framing because the continuous longitudinals make a bridge structure or girder of the hull, while the deep section web frames will resist athwartship strains far better than numerous shallow frames, and these are the reasons why, fifty years after CONSTITUTION, almost all scientifically built steamers are constructed this way.[1]

Said another way, because the primary loads on a boat are fore and aft, one wants as much material as possible running fore and aft to carry them.

The advantages of longitudinal framing for sailboats remain poorly recognized, perhaps because innovations in boatbuilding are little heralded. Usually, they are guarded as trade secrets to be used for commercial advantage. Not till after the demise, in 1946, of the Herreshoff Manufacturing Company did Francis Herreshoff break silence on the company's techniques with the publication of *Capt. Nat Herreshoff.* Further, making his way as a designer, not a builder, Francis publicized in *Common Sense of Yacht Design* his development of his father's steel longitudinal construction for building smaller boats in wood. His successful 1925 creation, the R-boat YANKEE, illustrates the technique.

As you can see in her construction plan, rather than the 36 commonly used close-spaced frames, YANKEE has 11 so-called "web" frames spaced 3 times farther apart. Running between them are 8 longitudinals, spaced about 12 inches apart, over which he constructed a shell of triple

---

1. *Capt. Nat Herreshoff*, p. 220.

CHAPTER 7 ⚓ THE STRINGER BUILT BOAT...

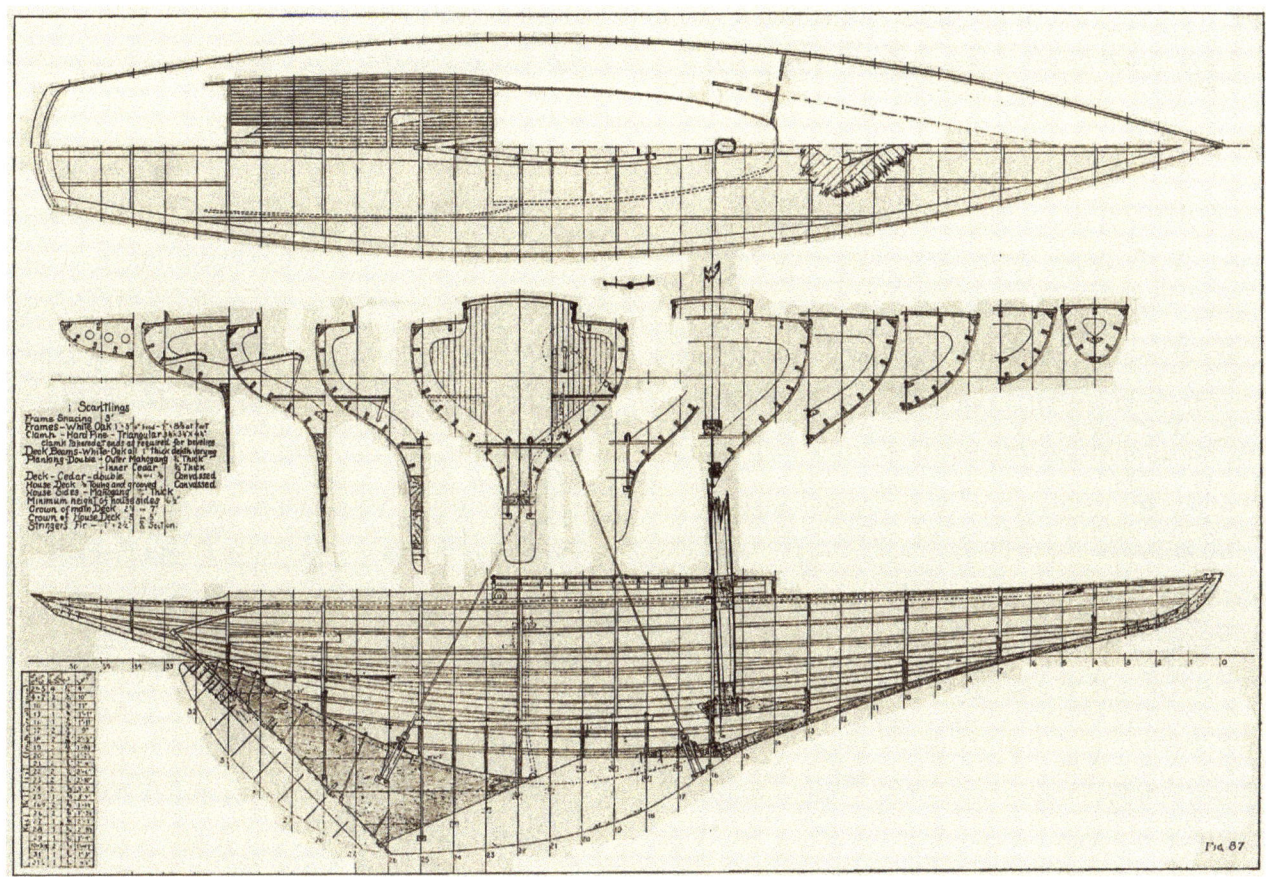

7-1 Longitudinal framing. *YANKEE*. Designed by L. Francis Herreshoff to the R class (under the Universal Rule). She is longitudinally framed with 11 web frames and 8 longitudinals each side. She is plated with Ashcroft planking, of which Herreshoff includes a sketch (just forward of the mast on the plan), similar to cold-molding, as in figure 5-3.

She must have been a hard boat to keep clean and dry, because the longitudinals trap dirt and water. Built in steel, she would be a rust trap. But she is strong and light, good for a racing boat. She had a very successful career.

diagonal "Ashcroft" planking.[2] Ashcroft's system, the direct forerunner of cold-molding before epoxy, used 3 diagonal layers held together with many small wood screws to form a bidirectional shell. Longitudinal framing requires bidirectional hull plating. Nathanael used metal for *CONSTITUTION*, and Francis used Ashcroft's wood for *YANKEE*. Once bidirectional hull plating is available, longitudinal framing, with the advantages Francis attributes to *CONSTITUTION* above, becomes hard to resist.

Besides being stronger, the longitudinal-generated hull has the advantage of being easier

---
2. The Ashcroft system of triple diagonal planking metal-fastened together was popularized in England in the 19th century.

to construct. It requires less formwork, less lofting, and less precision. Traditional transverse framing uses many frames. The many frames require precision in their lofting, cutting, and setup as the slightest variance from correct shape or position will generate unfairness. Traditional construction acknowledges this by using 10 or 12 molds that are set up and tied together with 10 or so temporary longitudinals called **ribbands**. The frames are bent within the ribbands. The ribbands give them both shape and fairness. The ribbands are then discarded as the planking is installed. Rather than using the ribbands as temporary forms to be discarded, the stringer-generated boat uses the ribbands (stringers) as structural elements that remain in the structure. While the traditional technique has the advantage of familiarity, it has the disadvantages not only of wasted ribbands but also that each of the many frames must be precisely shaped and then beveled to receive the planking. Precision and beveling are both expensive.

The stringer-generated hull

The stringer-built boat is a development of the longitudinally framed boat, with the wrinkle that the longitudinal stringers combine with the shell plating to form vierendeel trusses that stiffen it. The truss has a lower chord consisting of the stringer, a top chord of the hull plating and web formed by the cross section of light transverse frames. So stiffened, the hull does not require web frames.

This produces a hull similar to Nat Herreshoff's ceiled hull, except the interior of the hull surface is not concealed by the stringers (**RULE 3**). The overall thickness of the hull (plank thickness plus frame molding plus stringer molding) is about the same as Herreshoff's plank, frame, and ceiling. A day boat, lacking accommodation bulkheads, might require 2 or 3 web frames, especially if there is no deck.

Construction process for a stringer-generated hull

Longitudinal stringers, on the other hand, can be supported by as few as 6 molds, The stringers are placed over the molds, taking gentle and naturally smooth curves (cubic splines) that are easy to bend without use of steam or cutting. They need no beveling.

Construction of the stringer-generated boat begins with a mold that includes about 6 transverse bulkheads, sufficient in number to define the shape of the hull yet few enough to avoid the risk of generating unfairness. Over the molds, 6 or so longitudinal stringers are bent and fastened. Over these stringers, light, closely spaced frames are bent. The frames are of small sections. They require steamings but bend readily. Touching only the stringers and the hull plating,

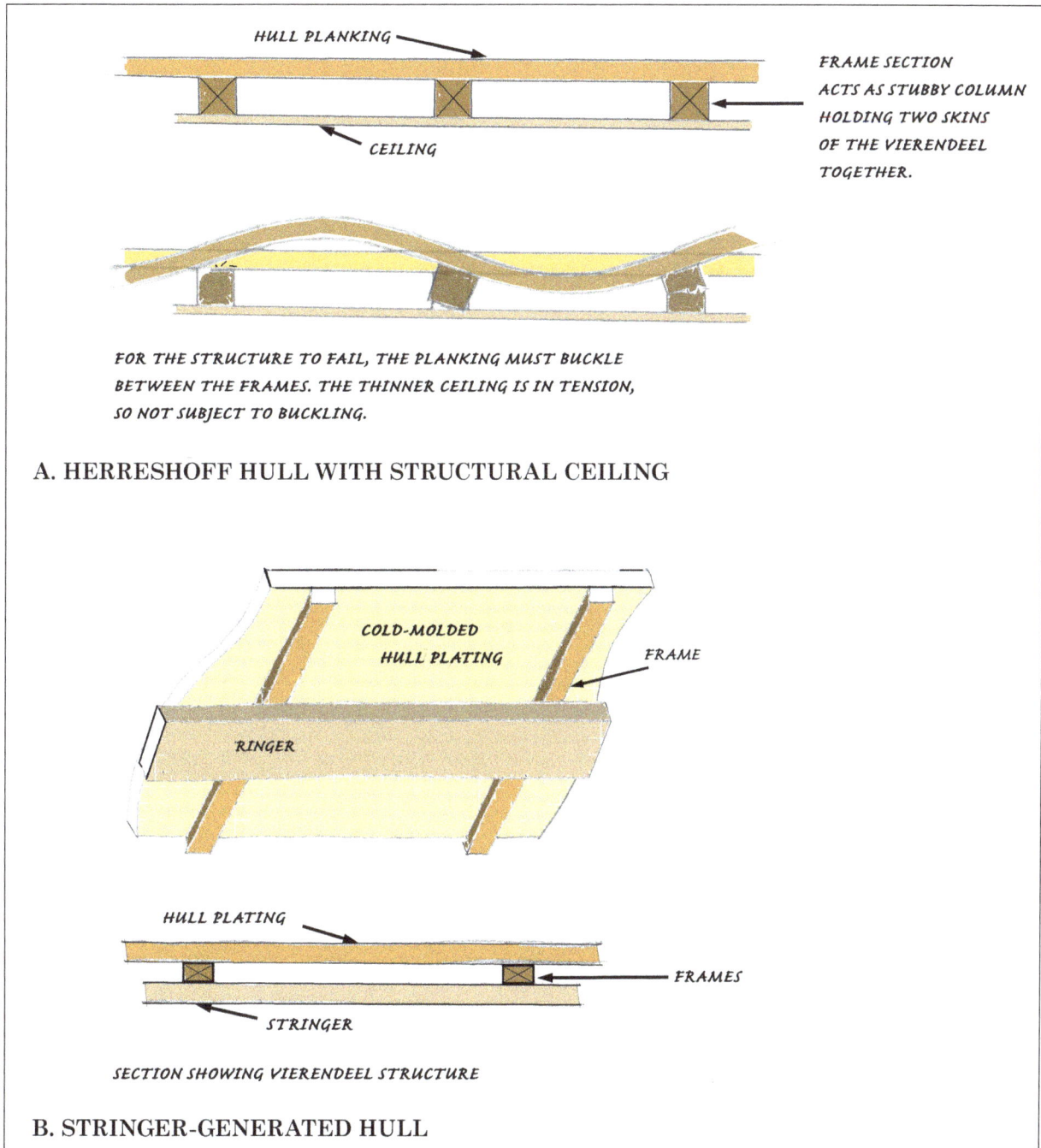

7-2 Herreshoff's structural ceiling. When N. G. Herreshoff made the ceiling of his boats continuous the length of the hull, rather than interrupted by the bulkheads, he created a sandwich structure that has about 10 times the stiffness of an unceiled shell. Herreshoff's two skins work together as a vierendeel truss. The top sketch shows how the truss would fail if the loading is great enough to pull the joints apart and buckle the planking. The bottom sketch shows Herreshoff's structure modified by concentrating the ceiling into larger, intermittent stringers, which allows access to the inner hull surface (**RULE 3**). This is used on the stringer-generated hull.

they need neither precise positioning nor beveling. They may cant as required to avoid need for twist. Quick and easy.

Structurally the frames act as the web of the vierendeel, separating the hull plating and stringers by their molded thickness. An important side effect of this separation of the stringer from the plating is that water and dirt fall unobstructed to the bilge. This eliminates the traps built into the seam batten hull we built for *FANCY*. Such traps are a nuisance on a wood boat and a danger to a steel one.

Over the light frames the hull plating is cold-molded. The interior accommodation is attached to the stringers, without touching the shell plating. Likewise, machinery and equipment attach to the stringers avoiding large point loads on the shell.

To summarize, the advantages of the stringer-built boat include

1. Easily generated shape, with correct number of transverse molds
2. Easily constructed hull shell. Simple parts
3. Entirely visible interior hull surface without water or dirt traps (**RULE 3**)
4. Interior accommodation securely connected to hull shell, without direct contact
5. Greater strength due to most structural material running in the direction of primary loads (**RULE 1**)

Comparison of different hull structures

A sailor loves his boat; most sailors love wooden boats the most. However, blue water sailors must be a practical lot; they trust their lives to their craft. They must loathe to go to sea with inferior strength. So let's look at how a stringer cold-molded hull compares to other constructions.

While there is no single number or even a small set of numbers that entirely describe the strength and resiliency of a sailboat, we can derive a couple of indexes that will give us a general idea of how different constructions compare. Basically, a sailboat hull is a shell that keeps the water out and carries the loads of wind and water. As discussed in chapter 4, a shell is very strong as long as it maintains its shape. But when deformed locally, it will fail in buckling. Local deformation is resisted by the local stiffness of the shell, which can be approximated with Timoshenko's formulas. Those formulae show stiffness is proportional to the modulus of elasticity (E) of the material used times the geometrical moment of inertia (I) of the section involved. We will also be concerned with bending strength, which will give us a proxy for resistance to puncture in collision or grounding. Bending strength is proportional to the breaking strength (σ_{max}) of the material used times the section modulus (S) of the section involved. We will use E*I as a proxy for stiffness and σ_{max}*S as a proxy for bending strength. Then we will examine a strip of hull shell running in the fore and aft direction for seven different constructions as they might be used on a 40 foot sailboat. These are the constructions:

1. Solid fiberglass
2. Foam-core sandwich fiberglass
3. Foam-core sandwich carbon laminate
4. Cold-molded wood, no stiffeners
5. Aluminum longitudinally framed
6. Steel longitudinally framed
7. Traditional plank on frame
8. Herreshoff plank on frame with structural ceiling
9. Stringer-generated, cold-molded wood

For all the designs, except #7 and #8, which are set by rule, I have manipulated the scantlings to give a shell weight of about 4.3 pounds per square foot.

Table 7-1 shows the results.

Description of the constructions

The first four constructions as they are used today produce a thin hull skin.

1. **Solid fiberglass** is the structure that started the modern plastic revolution. A gel coat, then alternating layers of mat and woven rovings. Thickness is ½ inch.

2. **Fiberglass sandwich.** Solid fiberglass, being strong but much too flexible, was first reinforced with stringers. Later, stringers were abandoned for sandwich construction. Two skins of glass-reinforced plastic were separated by a core of foam. The foam core must be strong enough to resist the horizontal shear in the sandwich. Horizontal shear increases with the tensile strength of the skins and the overall thickness. Offsetting this, a denser foam will have greater shear strength. The calculations assume 4 pounds per cubic foot foam, though lighter foams have been used. Skins are 7/32 inch separated by 1¼ inch of foam.

3. **Foam-core carbon fiber–reinforced epoxy plastic.** This is the same structure as #2 but uses the super-strong carbon fiber reinforcement. Because of the great tensile strength of carbon, the shearing forces are considerably greater than with fiberglass, so 12 pounds per cubic foam is included in the calculations. Skins are 3/16 inch thick separated by 1¼ inch of foam.

4. **Unreinforced cold-molded wood.** The simplest cold-molded hull consists only of the laminated skin without framing. It is similar in structure to solid fiberglass (#1) except much thicker. In this case it is 2⅜ inch thick Atlantic cedar. Cedar's density is 0.32 and fiberglass' is 1.69. As the density rule would suggest, the cedar hull is much stiffer and stronger than the fiberglass one.

TABLE 7-1 Relative strength of different constructions.

CONSTRUCTION TYPE			1 Fiberglass Solid	2 (note 1) Fiberglass/ Foam Sandwich	3 (note 1) Carbon/ Foam Sandwich	4 Cold-Molded No Frames	5 Aluminum Framed	6 Steel Framed	7 (note 2) Plank/ Frame Nevins	8 (note 3) Plank/Frame Herreshoff	9 (note 4) Cold-Molded Stringer Built
Item	Dimension	Units									
Plating	thickness	inches	0.50	0.22	0.19	2.50	0.25	0.08	1.06	0.96	1.13
Frames	molded	inches					3.00	3.00	1.50	1.68	1.00
	sided	inches					0.38	0.13	1.50	1.68	1.50
	spaced	inches					16.00	16.00	9.00	11.00	16.00
Stringer	molded	inches							1.00	0.42	1.25
	sided	inches							6.00	continuous	5.50
	spaced	inches							40.00	continuous	14.00
Foam	thickness	inches		1.25	1.25						
Total thickness		inches	0.50	1.69	1.63	2.50	3.25	3.08	3.56	3.06	3.38
Weight		lbs/ft^2	4.4	4.3	4.4	4.2	4.4	4.2	3.7	4.4	4.3

Relative to Herreshoff's rules (column 8):									
Weight	101%	98%	100%	96%	101%	97%	84%	100%	98%
Stiffness	1%	14%	59%	36%	59%	54%	55%	100%	106%
Flex strength	13%	86%	177%	113%	33%	17%	38%	100%	91%

High-performing structures highlighted in pink

Notes: See illustration 4-10, p.84
1. sandwich   panel "C"
2. plank on frame Nevins   panel "B"
3. plank on frame Herreshof   panel "A"
4. cold molded stringer built   panel "D"

The scantlings of the second four are manipulated to give the thickness similar to that given by Herreshoff's rules. Nos. 7 and 8 are the actual thicknesses required by their respective rules. Nevins is a little thinner than Herreshoff.

5. **Aluminum.** Metal skins reinforced by longitudinal stringers are a traditional alternative to the wooden boat. Aluminum is 7/32 inch thick supported by 3-inch deep stringers. The aluminum is stronger than steel, as expected from the density rule.

6. **Steel.** Same structure as aluminum. Steel is 14 gauge sheet (maybe too light to be practical) supported by 3-inch deep stringers.

7. **Traditional plank on frame** makes a beautiful, habitable boat. We use Nevins rules, which specify a single bilge stringer, and some guesswork is involved estimating its effectiveness in thickening the hull shell. Many Nevins boats have diagonal metal strapping that does not add to the shell stiffness or bending strength but will make them heavier. Frames are 1½ inch square white oak spaced 9 inches; the single bilge stringer is 1 inch molded × 6 inch sided fir.

8. **Herreshoff's plank on frame** adds a structural ceiling to the traditional construction in lieu of the bilge stringer. Planking is 15/16 inch thick fir, frames 1 11/16 inch square white oak spaced 11 inches, ceiling 7/16 inch thick fir.

9. Our much discussed **stringer-reinforced, cold-molded construction** is a derivative of Herreshoff's method, concentrating his continuous ceiling into a number of discrete stringers. Having a cold-molded skin, it uses lighter frames more widely spaced. The skin thickness is a bit thicker than Herreshoff's, the frames are lighter, 1 inch molded, 1½ inch sided, spaced 16 inches. The stringers are 1¼ inch thick by 5½ inches wide, spaced about 14 inches.

Results of analysis

Several results stand out from our analysis. Firstly, Herreshoff's rules produce the stiffest hull, beating the carbon sandwich hull almost 2 to 1. It is second in breaking strength at about 60% of the carbon sandwich. Stringer-generated cold-molding is next, 50% stiffer than carbon and half the breaking strength. The inadequacy of solid fiberglass is apparent and, although considerably improved by sandwich construction, still quite inferior. Steel and aluminum are better than fiberglass and inferior to wood and carbon. Aluminum is better than steel, as one would expect from the density rule. Carbon shows well but may not be worth its expense and brittleness.

The above analysis makes distinctions between different construction systems, but it does not answer the question raised by figure 4.10

of chapter 4 of how thick should a hull be. It may be that Herreshoff's construction, derived from tradition, required a thick hull because of the linear nature of his plating. It is possible that when bidirectional plating is used, the traditional thickness is greater than it needs to be. There are certainly modern boats sailing that have much thinner hulls. When all is said and done, the analysis indicates the stringer-generated boat is very strong and strong enough (**RULE 1**). Whether it is overbuilt remains, in my mind, an open question.

PART THREE

Building with Cold-Molding

Modern man fancies he can multiply his experiences, enjoy, understand and assimilate a hundred times more than his forbears. Instead, he is in the process of deadening his sensitivity, losing all perception of deeper things and changing into a mechanically reacting puppet, a machine which swiftly and superficially disposes of everything which gave people in former times a meaning in life. This is a tragic curse which we all seemingly favored moderns share.

> Goran Schildt, while sailing *DAPHNE* in the Dodecanese, 1957, *The Sea of Icarus*.

CHAPTER 8

Lofting

A BOAT IS CREATED in the physical world from a model. Models are used to define the form of the new craft and enable communication amongst her designer, builder and sailor. Models come in many forms. In more primitive shops the model may consist of wood templates for stem and bilge section. Herreshoff famously used precision half-models he carved in his studio. By the 20th century, most designers were using drawings to model a design.

Lofting, the first step of building a boat, is the process of extracting from the designer's model the exact shape of her parts. With pen and paper models this was done by drawing her lines at full size on the shop floor. The purpose is two-fold, first to smooth out any unfairness that might be found in the small drawings or offsets forwarded by the designer, and second to get the full size shapes.

Drawing the lines on the shop floor makes some sense because, usually, it is the only surface available big enough to draw the vessel at full size. But drawing big on the floor has its drawbacks. The greatest might be that such a large drawing is hard to see in its entirety. It is difficult to see unfairnesses, or even fairly gross mistakes, with the loftsman's eye only 5 feet above a 30 to 60-foot drawing. At such a flat perspective he can see clearly only a small portion of the whole. And drawing the lines thick enough to see from a distance means they will be at least an eighth inch wide, more than the tolerance for fine work.

When Sanford Boat lofted the *Alerion Class Sloop* in 1977, we tried an improvisation. Rather than layout the boat full size, I drew it quarter-size on mylar (3" = 1'). I found this had the advantage that I could stand back and see the whole boat at once as the drawing was only 7 feet long. Further, I found that on the fine surface of the mylar, my pencil line width and measuring ability had a resolution of about $1/64$ of an inch equivalent to $1/16$ of an inch at full size. Our building tolerance was $1/16$ of an inch, so this was accurate enough. We found a few (very few) discrepancies in the lines drawing from Mystic Seaport and the technique was quite successful.

By the 1980's, modeling was beginning to be radicalized by the computer revolution. Computer aided drafting (CAD) machines, which were moving out from the aerospace and automobile industry into more general use, generated a model in the form of numerical files. These files could be transmitted by email, depicted as drawings by printers, and easily manipulated by other computers. I had learned hand drawn graphics at MIT in the late 1960's from teachers who were, at the time, developing the first CAD machines. I dearly wanted one of these fabulous things. But when I called ComputerVision, the pioneer, back then, for small CAD systems, I discovered that a single station—computer, screen, tablet and a printer—would cost about $250,000. (Financing **IS** available!) At that time an *Alerion Class Sloop* retailed for $25,000, so that single workstation cost as much as our annual production. Back to the drawing board.

The end of the 20th century brought a thousand-fold drop in price and brought computerized lofting to the small boat shop. Designers, rather than forwarding small drawings or a tables of offsets to the builder, emailed numerical models of their hulls. From these computer files, all the necessary patterns could be precisely printed at full size.

In 2009, when *STARRY NIGHT* was to be laid down, the process was as follows: I drew the lines, scanned and emailed them to the Matt Smith who acted as our loftsman. He made a 3-D numerical model and emailed back a computer drawing of the building mold and its components. The builder and I approved the drawing, whence it was emailed to the laser cutting shop. A week later a truck showed up with a set of molds ready to use—a big change from the old days!

The loftsman, today, works with a computer, not with splines, nails, and pencils. Computer lofting, cutting, and marking has become commonplace because it increases accuracy and lowers costs. Computer lofting is as important to the reduction of the cost of building a wooden boat as cold-molding is to the increase

CHAPTER 8 — LOFTING

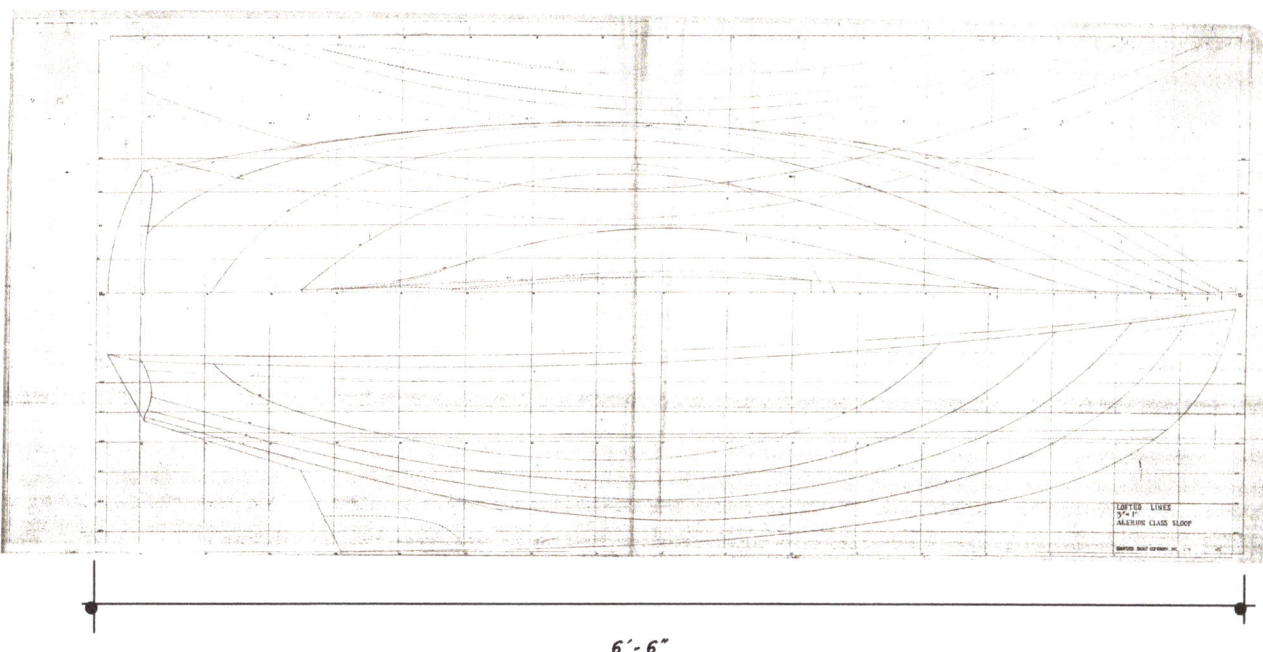

6′- 6″

8-1 Lofting less than full size. The *Alerion Class Sloop*'s lines lofted on mylar at quarter size. The drawing is 36 inches × 80 inches. The resolution of the drawing is $1/64$ inch, which at full size is $1/16$ inch and about the resolution of our construction ability.

of her durability. The cost reduction is greatest with a one-off design as all the cost reduction is applied to a single job. The cost reduction comes in two ways. First, the needed shapes are quickly generated by the computer. Second, two-dimensional pieces to be cut from sheet material can be cut precisely and quickly by computer-controlled cutting machines. It is almost free! The pieces coming from the cutting machine can also be marked with lines and writing, saving time and errors transferring information from drawings to the actual work in the shop.

Carefully used, computer lofting, marking, and cutting prevents layout errors. But one must watch out for gross errors. The danger of computer lofting is the human/machine interface. It is difficult to see both detail and overview simultaneously on computer screens—Francis Herreshoff had a 27 foot long drawing board for a reason.

The computer empowers the loftsman to be, more than ever, a crucial player on the builder's team; the computer has greatly increased his effectiveness and responsibility. He can be intimately integrated with the shop floor to accomplish much of the layout work that used to fall to the shipwright.

Curves of a boat hull

While there have been a multitude of hull shapes conceived and used over the ages, for the most

8-2 Splines and dogs. Naval architects used actual splines made of wood or plastic up till the CAD era. They were held in place by spline weights called "ducks." On the shop floor the splines were wood and held in place by nails put on either side of the battens.

part their shapes contain two distinct kinds of curves—the longitudinal curves along the direction in which the water flows and the transverse curves perpendicular to the water flow.

Curves parallel to the water flow want to be fair because fairness is believed to lower hydrodynamic drag. Curves perpendicular to the flow have little direct effect on drag but greatly affect other sailing qualities, such as stability, leeway resistance, and load-carrying ability.

Hydrodynamics wants the longitudinal curves that run with the flow of water to be fair and smooth—no sudden changes of direction or curvature—to ensure an easy flow with minimal drag. In the physical world, the longitudinal curves are formed by the planks of the hull. Planks, thin strips of wood bent over the shipwright's molds, take smooth curves. The designer models these shapes on paper in an analogous way, using thin flexible splines (made of wood or plastic) held in place by ducks. The positions where the weights hold the spline are called **control points**. The control points determine the shape of the curve.

Mathematicians studied these curves and learned that each segment between the weights (control points) can be represented by simple formula called a cubic polynominial (the formula is of the form $ax^3 + bx^2 + cx + d$.)

They called the overall curve a cubic spline curve. Analysis shows that cubic splines are indeed smooth. Besides looking smooth with no sharp corners or gaps, their slope and curvature is continuous across the control points. Being continuous in curvature is equivalent to no sudden accelerations of water particles along the spline which may be the reason for their easy passage through the water. I do not know; I have never seen this analyzed. But I have seen it stated that meandering rivers follow cubic spline tracks because those tracks minimize the energy required to move the water.

Cubic splines have the property that if the last control point is placed at the end of the curve, that the end will be flat. Designers often want some curvature at the end of their hulls, especially in the bow. This requires continuing the curve out into space with a control point

CHAPTER 8 ∞ LOFTING

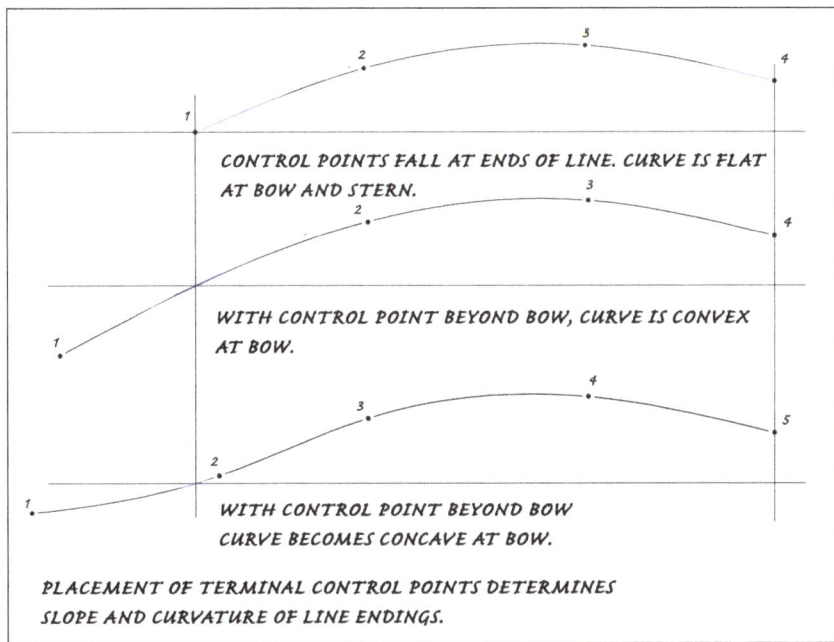

8-3 Spline end conditions. The curvature at the end of a cubic spline is controlled by a control point beyond the useful end of the curve. If the last control point is at the end of the curve, the curve will end flat. A control point beyond the end may introduce convexity or concavity.

beyond the end of the hull. As shown in figure 8-4 any curvature can be induced at the beginning of the curve, convex, concave or none. Of course, on the actual boat, you cannot hold a plank beyond its end. But curvature can be induced at the endpoints by manipulating the bevel on which the planks land. The bevel sets the slope of the curve at its beginning which in turn controls the curvature.

Transverse curves

The transverse curves, perpendicular to the flow, do not need to be fair, or even smooth. They are often "S" shaped, contain tight curvatures, and may even have sharp corners in the form of chines. Mathematically they are

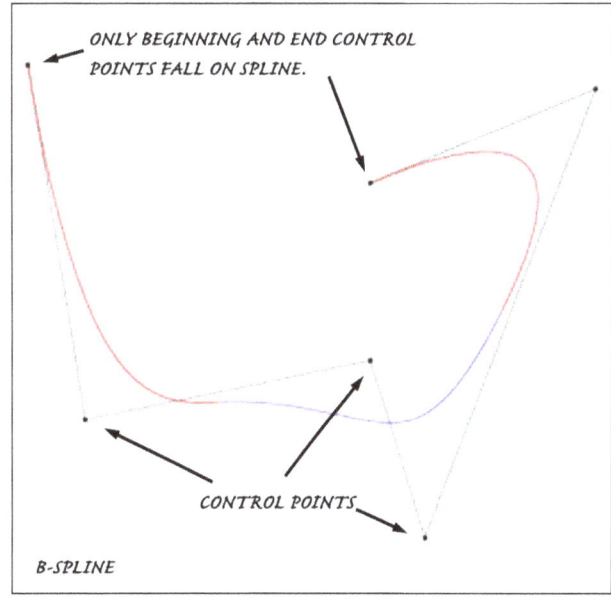

8-4 B-splines. Any curve may be generated by a b-spline. The b-spline was developed for computer design of aircraft and automobiles. B-splines are better for cross-sections than streamlines.

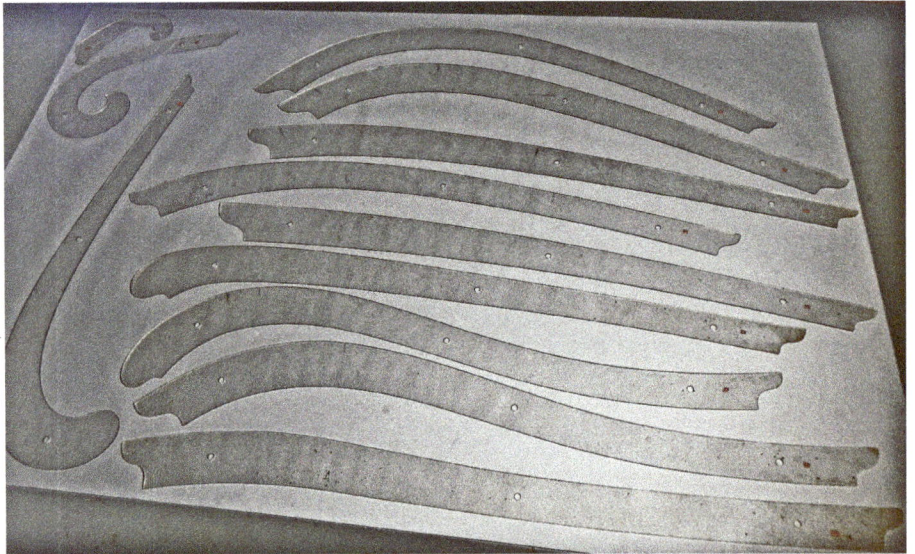

8-5 Ship curves. A partial set of Copenhagen Ship Curves. The numbers run to 150, so presumably that is how many in a complete set. These particular curves were made and sold by Eugene Deitzgen and Company in the 1950's, but the originals come from the 19th century and undoubtedly were made of wood. The long curves with little curvature are called sweeps and are cubic splines. The tighter curves are better represented by b-splines.

They are beautiful in themselves and, in the right hands, have produced extraordinarily beautiful sailboats. Their use has been obsoleted by the computer, so they no longer inspire the yacht designer. Their abandonment follows the general trend that has removed art from yacht design. The sailor suffers the loss.

represented by **b-splines**. On the drawing board they are drawn with straight edges, compasses, and ship curves. In the physical world they show up as frames. Their radically curved shapes are sawn from large pieces of stock or bent from very pliable steamed oak.

Now we live in modern times. Most sailboats today are designed using computerized drawing programs. These programs were developed for the automotive and aerospace industries and they render curves with the mathematical objects called b-splines, similar to but not the same as cubic splines. B-splines easily model traditional ship curves (figure 8-5).

Computer yacht design

One property of b-splines is that the control points do not lie on the curve they control and there is little intuitive relationship between where they do lie and how they control the curve. If you want a waterline 1 inch less beamy, it does not mean you move the control point 1 inch—it is, in fact, not clear how much you move the control point, or in what exact direction you move it. The process of moving the control point around to get the curve where you want requires doing the work iteratively: move the control point, look at the result, measure the result, move it again, etc. With feedback coming off a computer screen, the process takes some getting use to.

A new generation of yacht designers has learned to work with b-splines—they have become standard. B-splines render auto fenders and airplane wings very well. B-splines will render cubic splines if used correctly. The *Alerion Class Sloop* was lofted with real wood splines in 1977. In 2015 Matt Smith did it again by computer. The offsets generated by the two techniques were consistent within 1/16 inch, the resolution of the manual process. Modern designs appear, to some eyes, as lacking in variety and beauty. Whether that is contemporary style or a fault with b-spline, generated drawings, time, I suppose, will tell.

Fairness

Fairness is a somewhat subjective concept but suggests a smooth curve without waves or bumps. A curve may be fair without being the correct shape, but on a sailboat hull, the correct shape is almost always fair. Fair curves are believed to produce the least resistance to flow through the water.

Because of their mathematical properties, all cubic spline curves are smooth, but not all are fair; some are wavy or have hard spots. Generally, the fewer control points a spline has, the fairer it is. With too few control points one may not get the desired shape. With too many control points one risks waves and bumps (see figure 8-6). Once you have enough points to define the curve desired, usually about 6 on a conventional design, adding an extra point, if it is not in exactly the right place, risks unfairness. So the trick is to use enough control points to get the right shape and **no more**.

On the shop floor, the molds are the control points for the planks. The fairness issue remains. You need enough molds to get the desired shape but no more. With the precision of computer lofting, any number of molds can be generated. But to maintain fairness, they must be positioned exactly where they belong. A mold element has six degrees of freedom in space (three of rotation and three of position) and it is easy to get them a little out of alignment (see figure 9-3). A small number of molds

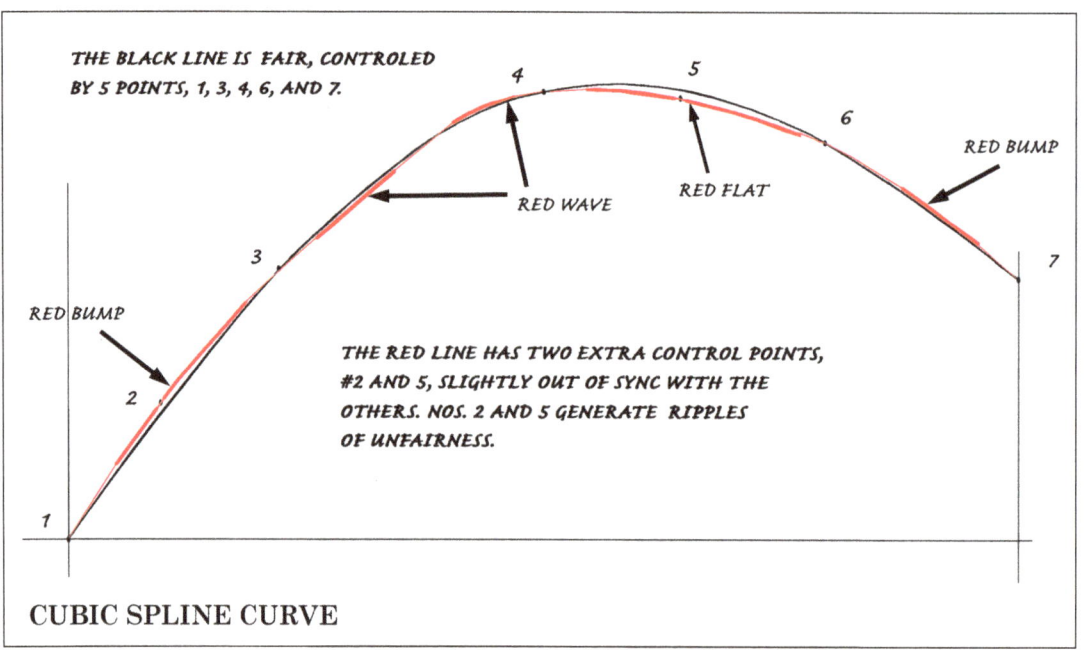

8-6 Cubic spline. This line might be a waterline for a sailboat. It has been exaggerated vertically for better visualization of the bumps. A cubic spline is sensitive to the number and placement of the weights. The weights on the drawing board are the mathematical control points. By moving the position of the control points along the curve, one generates (slightly) different curves. With too many control points—red line—one may introduce bumps and waves. The fairest cubic splines use the least number of control points to obtain the desired curve—black line. A control point's influence extends one segment beyond the adjacent segments. For instance, control point 5 influences both segment 3-4 and 6-7.

requires less shop floor precision. If the small number of molds are slightly out of position, the hull produced may differ slightly from the design, but it will not be unfair.

As we move on into the next chapter discussing the mold, let us remember:

1. The curves along the lines of water flow are fundamentally different from the curves of the sections across the water flow, in mathematics, on paper in the shop, and in the water itself.
2. Too many control points may make a curve unfair.
3. Computer controlled cutting of sheet material has a big place in wooden boatbuilding.

CHAPTER **9**

The Mold

ACTUAL CONSTRUCTION BEGINS with the building of the mold. The mold both shapes the boat and organizes its building process. The mold has never much affected the interior accommodation of the boat. But I think it should. I'll tell you the story of my gradual realization that the mold might be the answer to the question that had been haunting me since the *Alerion Class Sloop*—how to quickly and easily install cabin accommodation into an oceangoing sailboat.

For four decades, I had been thinking about techniques for installing a prefabricated interior into a completed sailboat hull. I had discovered the difficulties while building the *Alerion Class Sloop*. She was a very fine piece of work, and her interior joints were exposed to view. With her, we just toughed it out and hand-fit her bulkheads to the curve and bevel of her hull. This required great precision and craftsmanship—both are expensive.

With *FANCY* came the idea to build the interior accommodation first and then to construct the hull around it. This worked pretty well, and solved the fit problem, but had some drawbacks. The rough work building the hull abused the highly finished accommodation pieces. The technique also required building the hull right-side up, which is more difficult than building upside down. What was gained with the accommodation was lost with the hull finishing.

STARRY NIGHT was stringer built. Her structure minimized the interaction between hull and accommodation. The accommodation pieces landed on the stringers and did not touch the hull surface. This meant there were only point connections to make, via the bulkhead blocks described in chapter 3, so the pieces were easier to fit than those of the *Alerion Class Sloop*. But it was still not easy to get the pieces lined up correctly, both with each other and in the hull itself. If one piece were installed out of place, it would lead all the other pieces joined to it out of place, making a mess of the whole job.

I'll admit that I am not very fast, and there were false starts. I had retreated from the box canyon of *FANCY* and made some progress with *STARRY NIGHT*, but an elegant technique for locating, fitting, and fastening accommodation pieces eluded me. My intuition, or whatever it is that haunts the mind at 3 AM, led me to think that the desired technique involved the mold. *FANCY* used the accommodation for the mold, but I knew that was not the final answer. My thinking remained fuzzy and the thought of a special mold would not go away.

The mold not only controls the shape of the boat, it is also tightly integrated with the building process. Early small-craft builders used a simple mold made of the project's centerline structure set up on the shop floor with three or four station-molds set upon it to establish the three-dimensional shape.[1] Longitudinal ribbands were bent around the station-molds from stem to stern. The ribbands acted as splines, and the stem, station-molds, and transom were their control points. The control points being few, the resulting shape was fair and insensitive to precision. The technique was economical. Frames were bent inside the ribbands. The ribbands shaped the frames and kept them fair with each other. As the planking was fastened over the frames the ribands were removed and discarded.

Herreshoff developed quite a different technique. He designed his boats using precision wood models. He invented an instrument to take off the offsets for each frame. He sent these offsets to the shop, where his foremen made a frame-mold for each frame. Sometimes the frame-molds were set up on the shop floor to have the frames bent over them. Sometimes the frames were bent over the forms on the bench and were then set up on the shop floor, beveled and planked. I am not sure how the complex was kept fair. The hull shape would have been massively over-controlled. Probably the frames were set up and then adjusted with battens to show the bevel and proper positioning.

---

1. Station-molds take their shape from the cross section of the hull at a particular station (position) along the hull's length.

CHAPTER 9 — THE MOLD

Herreshoff's techniques were not widely copied. After him most fine construction used scale drawings. The mold was made from the lines drawing. Commonly, there might be ten stations spaced evenly along the waterline. Station-molds are cut to the shape of the hull's cross section at each station. The mold might have 12 or so elements, one for each station and, perhaps, extra ones in the ends to better form the bow and stern. There were sufficient stations to risk unfairness from over-control, so care had to be taken. Ribbands were installed. The frames, of which there might be 50 to 60, were bent into them. The planking proceeded in the usual way and the ribbands were discarded.

STARRY NIGHT's mold was a station-mold type with computer lofted and cut station-molds spaced 3 feet apart (see figure 3-10). The mold held the centerline structure in place and defined her sheerline. The stringers, which generated her shape, were bent directly over the mold. The rest of the hull was added to the stringers.

Each of these mold techniques leaves the builder with an empty, upturned hull into which the interior accommodation is to be installed.

Traditionally the accommodation was built piece by piece, in situ. The shipwright was able to correct any errors of placement as he went. If a bulkhead were off station ¼ inch, he could make the adjoining panel front ¼ inch longer or shorter to compensate. No one would notice the misalignment, and if they did, they would not

9-1 The bowl. *STARRY NIGHT* uprighted with the mold removed. She is a beautiful thing, but hard to get in and out of, or even to stand up in. The question remains, where do the bulkheads go?

YELLOW ARROWS POINT OUT COMPLEX JOINTS REQUIRING NEAR PERFECTION.

9-2 Where to start? The sole framing being installed in *STARRY NIGHT*. The overall frame had to be planar flat fore & aft and athwartships and at the correct height. The job took a master shipwright 2 weeks to accomplish.

care. The interior would fit together beautifully and functionally.

But the process is slow going. The working conditions inside the bowl include cramped space, containment, and difficult access. To use machinery, the shipwright must climb out of the boat carrying his work to the machine, do his job, and then bring it all back for another fitting. The curved surfaces inside a forty-footer are hard to stand in, much less to measure or locate position accurately. With *STARRY NIGHT* it took a master shipwright 8 days to build the cabin sole, flat, level on two axes, and at the correct height.

Economy and quality want the accommodation parts to be prefabricated on the bench and prefinished in the paint room. Such prefabrication, while saving time and increasing quality, demands precision. Once made, parts are not easily changed. They have fixed relationships to the neighboring parts built into them. Each part enjoys six degrees of freedom, three of position, and three of twist, And if slightly out of place or orientation, not only will it misfit the hull, but, more importantly, it will be upset in its relationships with its adjacent pieces. Changing one piece demands changes of them all. Furniture tolerances are about $1/32$ of an inch, so even a small misfit usually means the piece must be discarded and a new one built.

The *Alerion Class Sloop* introduced the difficulty of fitting an interior, and both *FANCY* and *STARRY NIGHT* teased my thinking about it. It was the sort of puzzle that wakes you up at 3 AM and sets you thinking. The puzzle reduced to three questions:

1. Where, exactly, was a particular accommodation piece to go?
2. The position found, what is the precise shape of the hull there?
3. How to prebuild the piece to fit that shape?

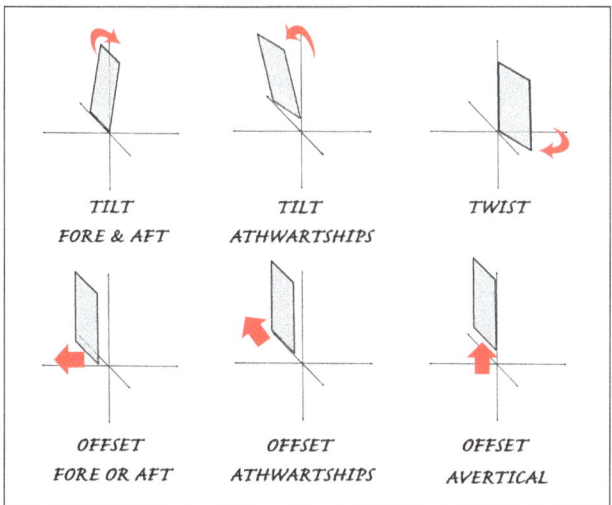

9-3 Six degrees of freedom. A bulkhead installed into the hull shell can be out of place in six different ways, three of position and three of direction.

But it was gradually becoming apparent to me that the right kind of a mold might just answer these questions. With *STARRY NIGHT*'s stringer construction, the three questions were reduced to two:

1. Where do the bulkhead blocks go?
2. What is the shape of the hull at their location?

One 3 AM, the insight I had been longing for arrived. If the mold was structured by the accommodation plan, rather than being structured by the lines drawing, the answers to the questions above became trivial. Let us call it an **accommodation mold**. Make a mold element for each accommodation piece. Where Herreshoff used a mold [element] for each frame, let us use a mold element for each accommodation piece.

Question #1 is answered by attaching the bulkhead blocks to the bulkhead-molds before the hull is even constructed. Question #2 is answered by the shape of the bulkhead-mold itself. Use the bulkhead-mold shape again for the actual accommodation bulkhead. How simple! In a way it is what we did with *FANCY*, but instead of using the actual finished accommodation pieces, we will use rough, temporary pieces, cut with computer-driven machinery, that stand in for the accommodation pieces. After the hull is turned upright, we will exchange the mold elements with identically shaped accommodation pieces.

Some terminology

Before describing the accommodation mold in detail, we will need to define some special terms to speak clearly about the new structure. The term "mold" has been used indistinguishably for both what I call a "mold element" and the whole thing. In this book I call the individual pieces **mold elements** and the entire structure the **mold**. Accommodation pieces need terminology also. Accommodation pieces are usually planar and run in three different directions. The **bulkheads**, a traditional term, are vertical planes running athwartships, across the boat. Vertical pieces running fore & aft do not have a traditional name; I call them **panels**. Finally, for horizontal surfaces, such as seat bottoms, bunks, counter tops, I drop the "horizontal" and just call them **surfaces**.

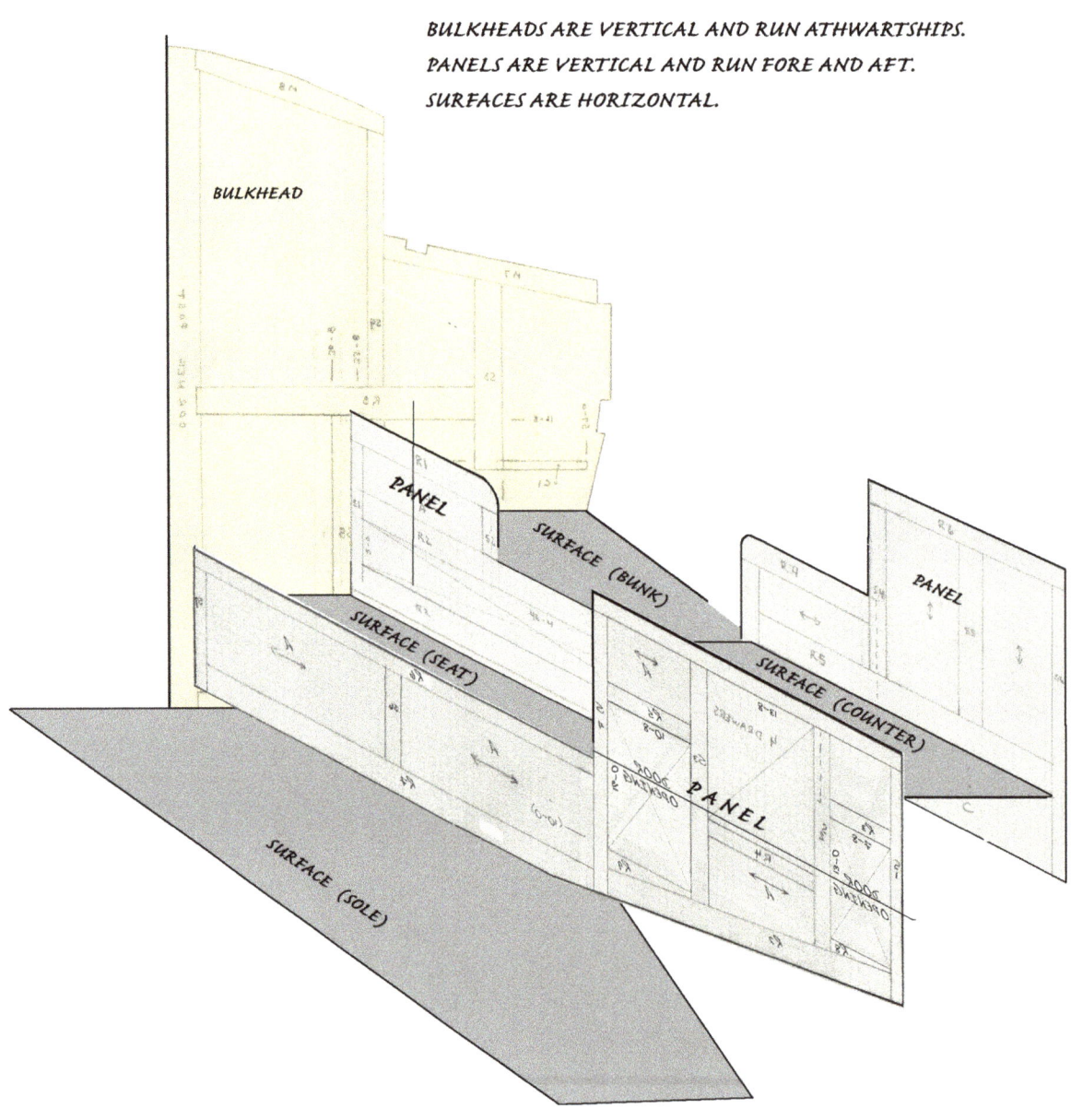

9-4 Terminology. "Bulkheads" are vertical and athwartship, "panels" are vertical and fore & aft, "surfaces" are horizontal.

When talking of the mold elements used as stand-ins for the accommodation pieces, I use the same terminology suffixed by "mold," hence **bulkhead-mold**, **panel-mold**, and **surface-mold**. The **sole-mold** is a special surface-mold, standing in for the cabin sole. Where there is no actual bulkhead to provide the support which the stringers need, we may add a temporary transverse, vertical mold element unassociated with an accommodation piece, which I call a **station-mold**.

Building an accommodation mold

Mold building is facilitated by the recently developed technology computer-controlled laser cutting of plywood panels. This is particularly true for building an accommodation mold. After the loftsman has designed each piece of the accommodation, at little cost his files can be cut accurately and repeatedly. Two reliably identical patterns can be cut, one for the mold, sent to the shop floor, and the other for the accommodation piece, sent to the cabinet shop. Both will be accurate to the 1/32 inch tolerance required. Intricate shapes can be cut, such as jigsaw puzzle joints to join several pieces to make a large panel, and holes can be drilled.

Not as well recognized, the laser can also mark the stock, so layout lines, joint locations, and even building instructions can be marked on the stock. This brings the design directly to the shop floor, eliminating the garbling of message as it passes down the chain of command. It also eliminates a lot of field measurement, which is the source of many errors. The old saying "Measure twice, cut once" is replaced by letting the machine do both the measuring and the cutting.

The mold begins by fastening the mold elements together, upside down on the shop floor,[2] because we wish to build the hull upside down. The panel-molds and the surface-molds will hold the bulkhead-molds in their correct positions and orientations.

The mold beyond the accommodation

The fore and aft peak of most designs will not have sufficient accommodation to form the hull. In these regions we will revert to the common section-molds. At the bow there will be a fore & aft mold element along the centerline to position the stem. In many designs one will want an extra section-mold near the bow to form initial curvature in the stringers.

In the aft peak, most designs will have a centerline fore & aft mold element running from the aft end of the sole-mold to the transom. This will support the deadwood, any skegs, the horn

---

2. For the rest of this chapter I will use "up" and "down" in reference to the upside-down mold. Since the mold is upside down on the shop floor, this will be the reverse of the actuality on the finished boat. So in this chapter the keel will be on top and the deck on the bottom.

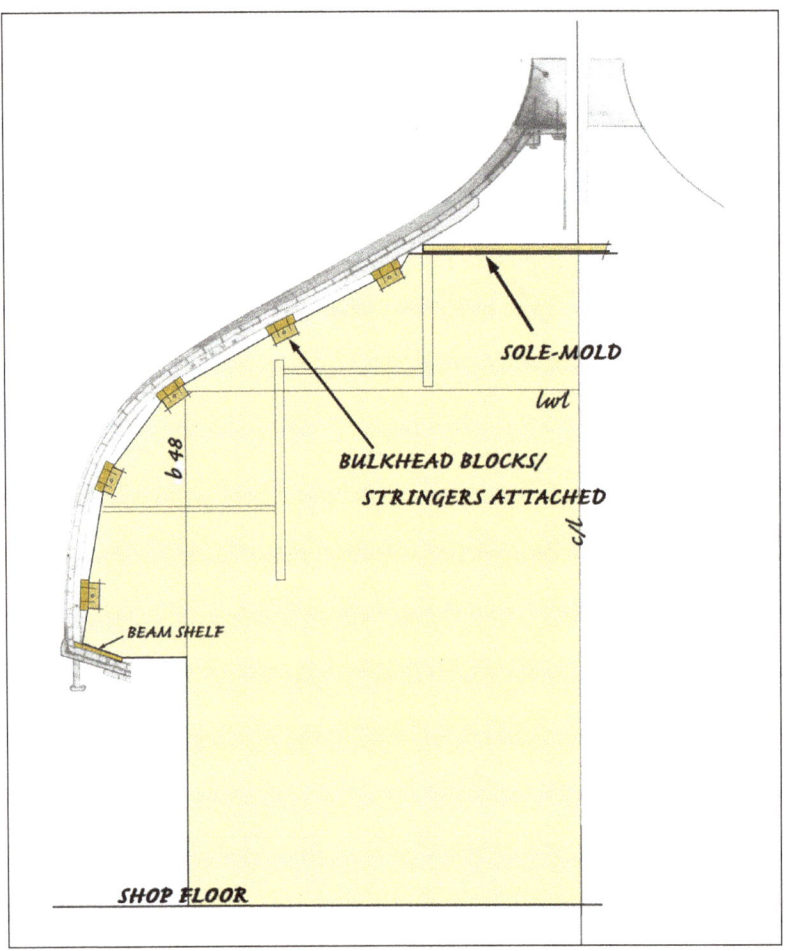

9-5 Typical bulkhead-mold. The bulkhead-molds hold the stringers in position and locate the bulkhead blocks. The sole-mold sits atop it.

deck. It leaves the bilge undefined, which is just as well, because the bilge is a whole different world, a sump for debris and water and the space where the hull shell joins the keel. Although the bilge needs its own mold method, the mold elements still have the same two purposes as elsewhere, to shape the hull surface and to locate the defining interior elements. In particular, the frames must be terminated, and the sole beams must be located.

Each keel type—long, short, fin, etc.—will need the bilge molded to carry the particular load of that type of keel. Figure 9-7 shows the bilge mold for a long-keel, shoal-draft hull similar to STARRY NIGHT's. It uses a section-mold to terminate each frame and to shape the laminate below the frame ends.

timber, and the transom. It will tie in with the station-molds forming the aftpeak. Usually, at the stern the stringers will terminate with flat curvature and will need only a simple landing for their ends.

Molding the bilge

The mold, as we have described it so far, molds only the hull in between the cabin sole and

The mold in the bilge begins with the sole-mold. The sole-mold, the stand-in for the cabin sole and a piece not found in traditional mold, separates the cabin space from the bilge. I not only prefer to support the panels and surfaces from the bulkheads, I believe it best to land the bulkheads on the sole. This makes the lower side of the sole-mold the base for the bulkhead-molds.

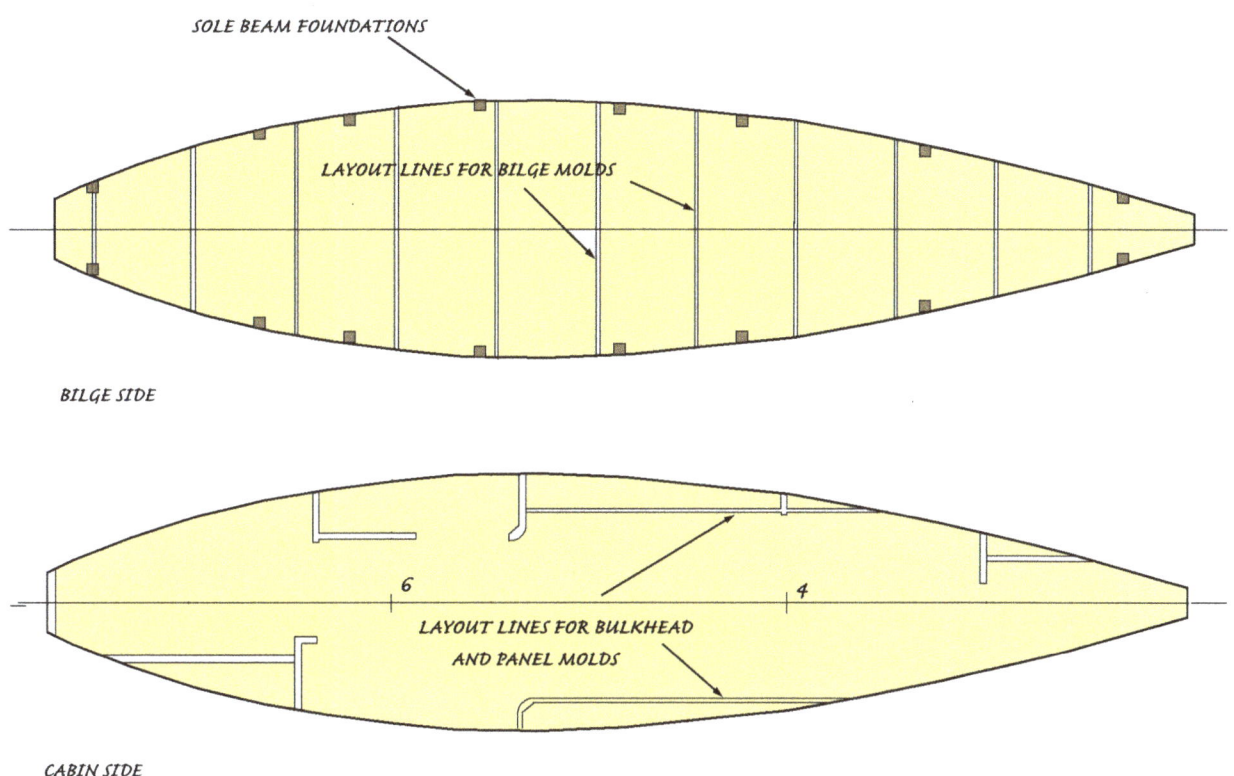

9-6 Sole-mold. The sole-mold separates the cabin from the bilge. Importantly, it carries the foundations for the sole beams. The sole-mold is a new piece not found in traditional molds.

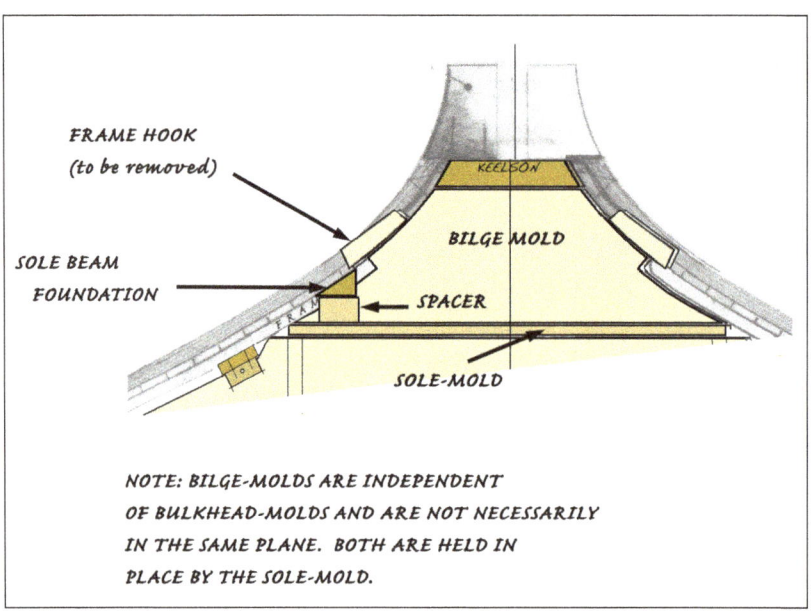

9-7 **Bilge-mold**. The bilge-molds sit atop the sole-mold. They provide a landing for the bottom of the frames and support the hull plating between the frame ends and the keelson.

The top of the sole-mold forms a platform upon which the sole beam foundations and the bilge-molds are placed. The sole beams, foundations are attached temporarily atop spacers whose height equals the sole beam depth. The foundations will be glued to the plating as the hull is laid up and thus awaiting installation of the sole beams when the hull is righted. Notice that they occur where the cabin sole requires them for its support; they rarely coincide with the bilge-molds.

The bilge-molds are erected on the sole-mold. They occur at intervals to terminate the frame ends and support the laminate between the frame ends and the keel. The keel will be laid atop them.

Here is the mold complete, now ready for the assembly of the hull and, after hull completion and turnover, the installation of the interior.

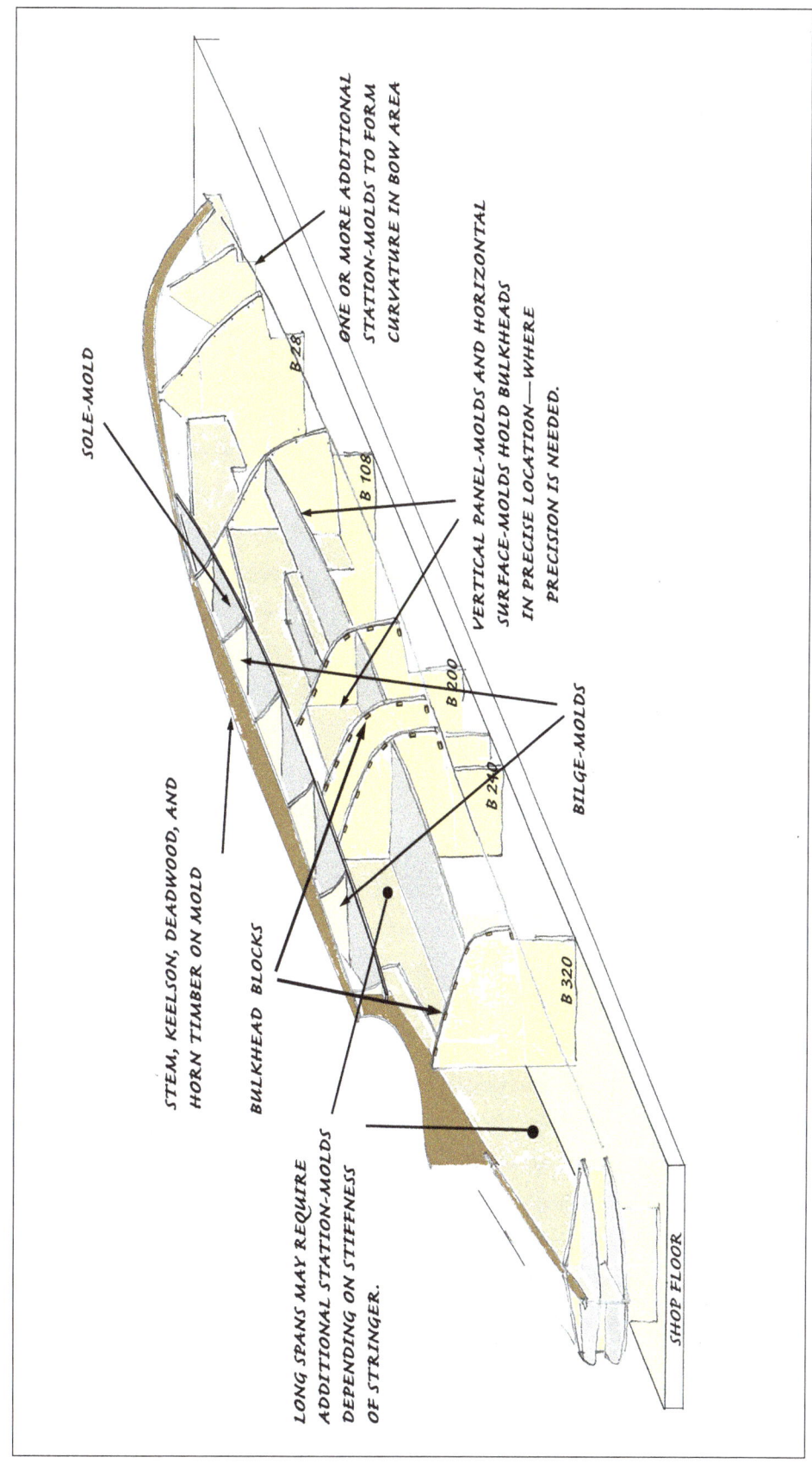

9-8 The completed mold. Using the accommodation as a mold locates the bulkhead blocks, giving precision just where it is needed. It makes installation of the interior a trivial job.

CHAPTER **10**

Building the Hull

COLD-MOLDED, EPOXY-BONDED boatbuilding is a new art. Where traditional wood construction has developed in Europe and America for at least a thousand years and been used on millions of vessels, cold-molding is about 50 years old and has been used, my guess, on a few hundred specialty sailboats. As I have mentioned, new inventions take three tries to get right and this chapter presents my third try, with *FANCY* being the first and *STARRY NIGHT* being the second. The boat described has not actually been built, think of the chapter as a thought experiment. Call the process the Sanford method, if you will. It is a method that uses a stringer generated hull, prefers bent lineal stock to 3-D shaped pieces, and can be used in a traditional shop using simple tools, but does require computer lofting and cutting of sheet material.

Building a cold-molded hull is similar to, but different from, a traditional build. The switch from metal fasteners to epoxy bonding changes each piece that goes into the construction, in both shape and assembly. By making the changes thoughtfully and exploiting the benefits of epoxy bondings we can build a boat without most of

CHAPTER 10 — BUILDING THE HULL

the water traps, rot generators, and weak spots found in traditional construction.

We described the new mold in the previous chapter. The process of building begins with setting up the edges of the hull shell. The beam shelf establishes the sheerline; the centerline structure—stem, keel, deadwood, horn timber—sets the lower edge of the hull, and the transom terminates the hull shell in the stern. Next, using the bulkhead blocks, we will add the stringers to define the intervening hull surface between sheer and keel. Over the stringers we will bend light frames, which we will then plate with a cold-molded skin. This chapter will describe the process in detail and describe the new shapes of some of the key parts that derive from epoxy bonding.

Beam shelf and sheer clamp

Traditionally, the beam shelf and sheer clamp are two heavy timbers running fore & aft the

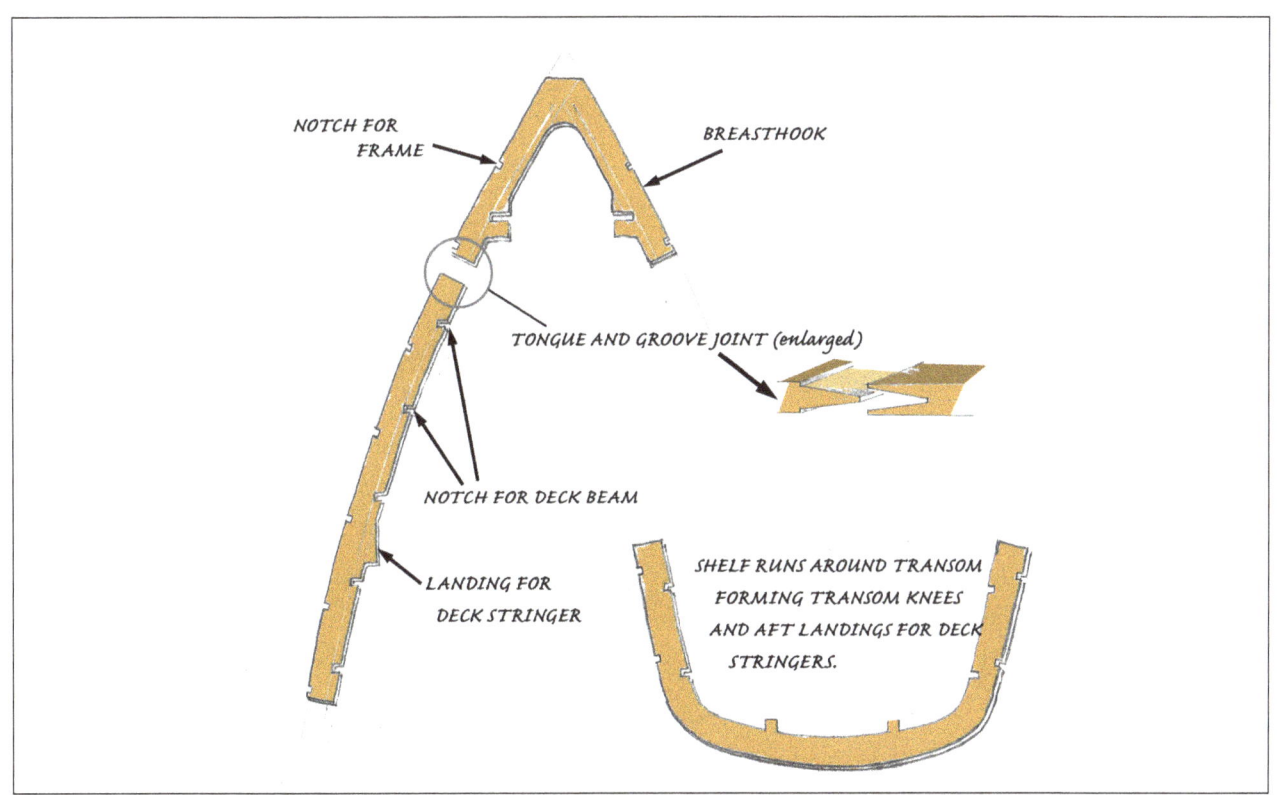

10-1 Beam shelf. The beam shelf is a plywood band running around the entire hull. It is reinforced by a narrow stringer on its inboard end and is notched to take frames ends and deck beam ends. It also provides landing for the deck stringers. The beam a shelf is an example of a complex shape easily produced by computer controlled laser cutting.

length of the hull at the top of the hull and at the edge of the deck (see figure 13-2). The clamp is vertical and bolted to the frames. The shelf is horizontal and bolted to the deck beams. The two are bolted to each other, tying hull and deck together. With cold-molding, the structure takes a very different form. The clamp almost disappears and the beam shelf becomes a plywood sheet reinforced with a narrow stringer. The shelf stringer, narrow enough to bend to the horizontal curve of the shear, is attached to the bulkhead-molds with bulkhead blocks. On a forty-footer it will be about 6 inches in from the sheer. Then the shelf proper, a plywood band spanning from the sheer to the stringer, is glued to the stringer. The shelf is made up in plywood lengths joined end to end with glued, tongue and groove joints. It runs right around the entire perimeter of the boat. It is notched to receive (and locate) the frames on its outboard edge and notched again for the deck beams on its inboard edge. There are protrusions on the inboard edge to land the ends of the deck stringers. Forward, at the bow, the shelf goes right around the stem

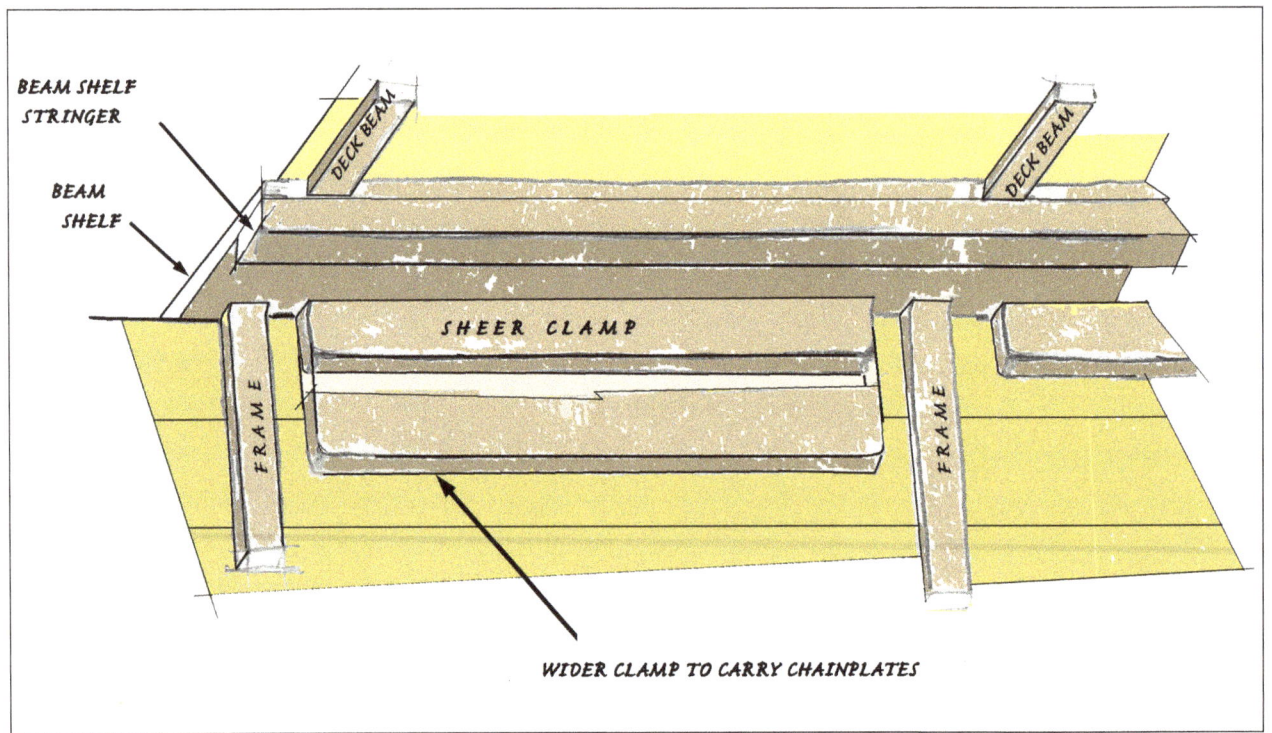

10-2 Sheer clamp. On a cold-molded hull the sheer clamp becomes an intermittent block reinforcing the hull/deck joint. It will transfer large tension loads into the hull shell. Its size is determined by the size of the loads. The sketch shows a smaller size suitable for supporting a rail chock and a larger size that might be suitable for a chainplate.

forming the breasthook. Similarly, aft, it runs around the transom forming transom knees, the transom frame, and deck stringer landings as it goes. It is a complicated shape well suited to computer lofting and laser cutting. The beam shelf defines the sheerline and will serve as the landing for the deck plating.

Meanwhile, the traditional sheer clamp becomes a vestigial piece, taking the form of short blocks used intermittently between some of the frames. It forms a reinforcement to the hull where large vertical loads, such as chainplates, genoa tracks, and rail chocks, are to be transferred into the shell. It is glued to the beam shelf between the frames. As the hull is planked up, it will be glued to the hull planks. The sheer clamp sections require a bevel on their top edge, cut to the angle between hull and deck. The bevel varies along the length of the hull, being acute forward and aft while slightly obtuse amidships. With the traditional clamp and shelf this bevel requires careful marking and cutting of both pieces for their entire length. But the cold-molded clamp is easier. It is made of a series of short pieces. Each has a fixed bevel that is easily cut on the table saw.

Stem

The traditional stem has a rectangular cross section cut with a rabbet to receive the forward (**hood**) ends of the planking. This rabbet is a complex cut, curved in profile to the shape of the bow and continuously varying in bevel as each plank lands at a different angle.

The traditional stem projects forward of the rabbet enough to allow sufficient thickness under the faying surface of the planks to hold the plank fastenings. The stem protects the hood ends of the planks, which are vulnerable both to water entry and to being peeled back in any sort of collision. After planking, the traditional yacht stem is usually carved to bring the outer surface to a smooth conclusion and may or may not have a stem shoe to protect the pointy end from minor collisions.

Cold-molding simplifies the stem joinery because the planks will be glued to it. No stem thickness is required to hold fastenings. All that is needed is a broad enough landing for a good glue joint. Consequently, the stem ends in a point at the rabbet line. After planking and before the laminate is applied, fiberglass tape (more about this later) is laminated around the stem, which seals the hood ends and ties the two sides of the hull together. After the veneers of the laminate are in place, they are cut to the bow profile and covered by a glued-on stem shoe. This gives the effect of the traditional rabbet without the need for cutting it.

The stem shoe is made of two thin layers. The first layer is screwed into the stem from ahead and is wide enough to project far enough beyond the hull surface that it may be grasped by C-clamps. When the glue dries, the screws are removed and the second layer is glued to

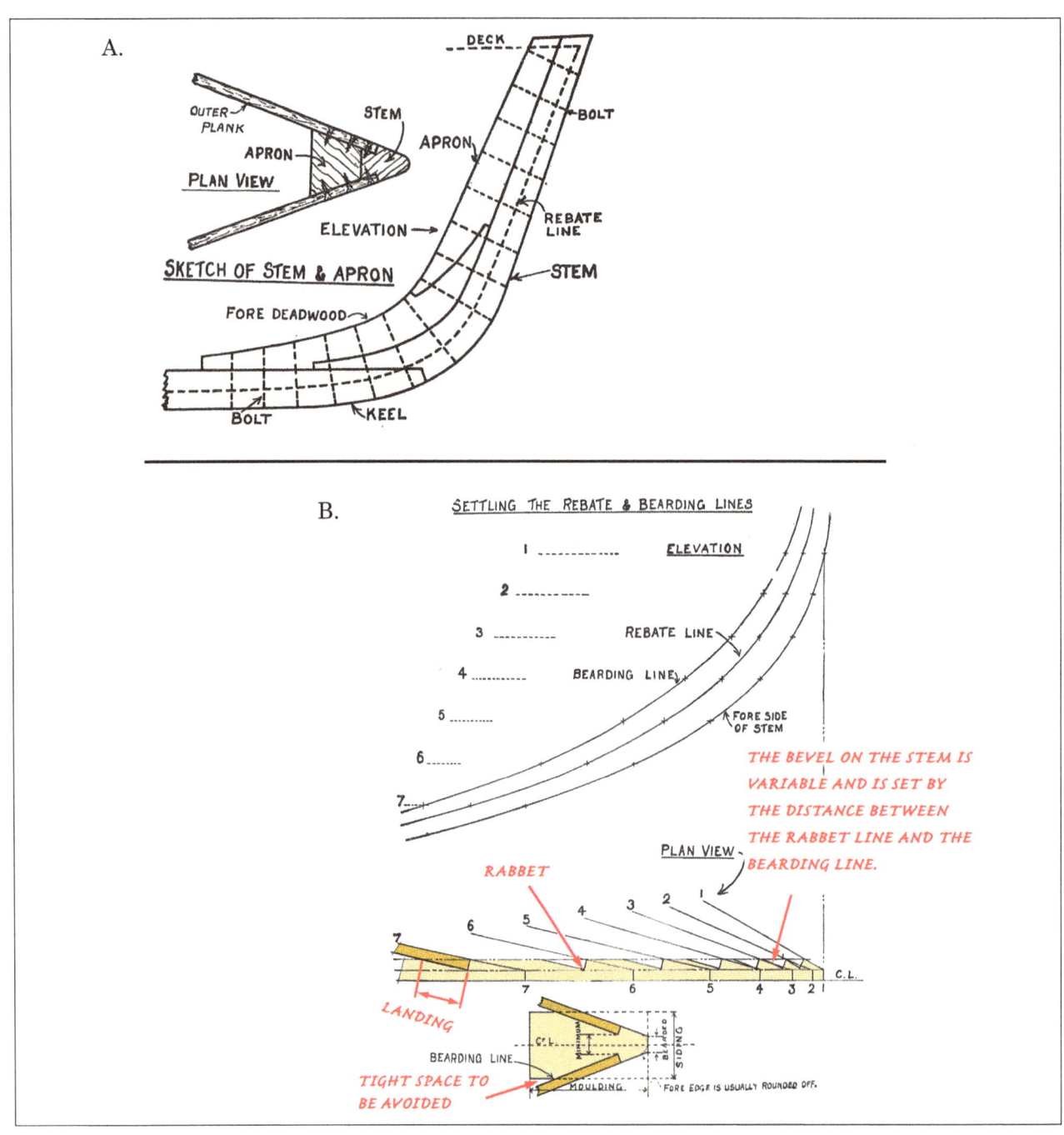

10-3 Traditional stem. This drawing is reproduced from C. W. Watt's 1947 manual. Watts work for Camper and Nicholson and his design, "A," may be taken as state of the art for traditionally built English boats. "B" shows the lofting that obtains the rabbet line and bearding lines.

If the stem continues aft of the bearding line as it does in "B," tight pockets are created that trap dirt and moisture. In "A" the aft edge of the apron coincides with the bearding line, eliminating the pockets. This is a detail that deserves to be copied. (C. W. Watts, p. 20).

CHAPTER 10 — BUILDING THE HULL

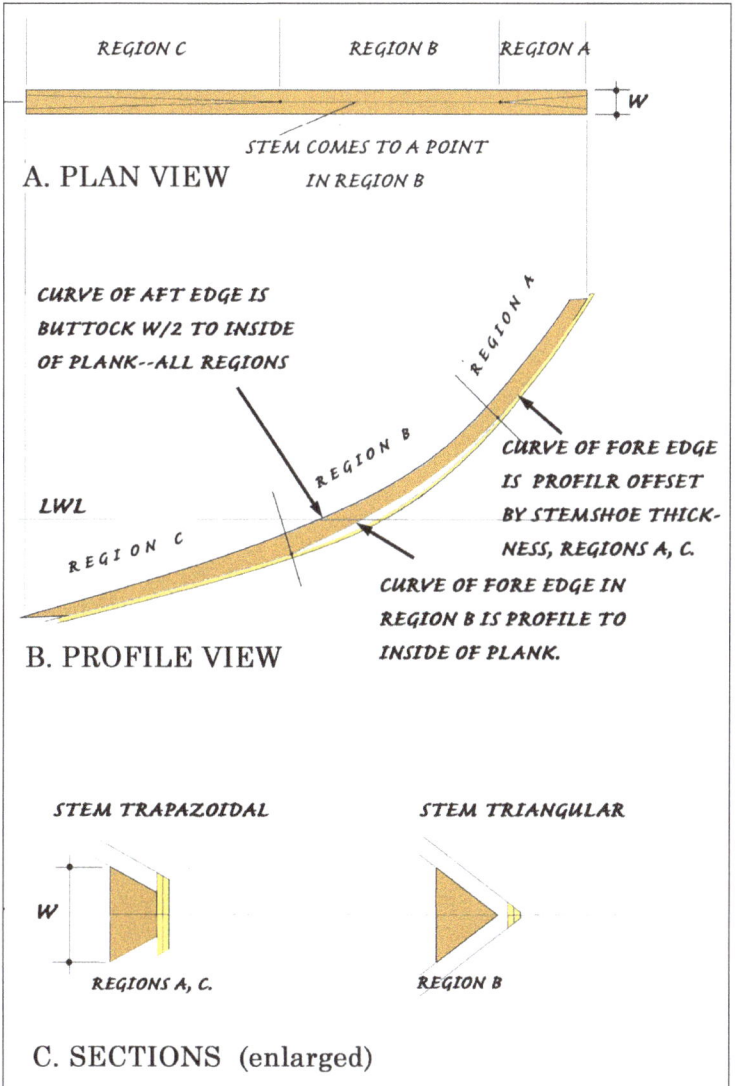

10-4 Cold-molded stem. The cold-molded stem does not need thickness for fastenings and we will terminate its aft side on the bearding line. This sketch shows the derivation of its shape. The loftsman does the layout, leaving the shipwright a fairly simple job. The wright saws out the profile and forms the continuously changing bevel by cutting off the corner with his plane.

10-5 The stacked stem of the *Alerion Class Sloop*. Here three stems are being constructed. The pieces have been cut to their contours, per figure 10-4. After gluing up, the corners will be planed off to give the triangular/trapezoidal cross section. Note, for the *Alerion Class Sloop* we cut the profile for each layer of the stack, which reduced the amount of planning required.

the first. being held to it with the clamps. When that glue has cured, the clamps are removed and both layers are trimmed to the hull shape with saw and plane.

The cold-molded stem will be triangular in cross section or, where the landing angle is broad, trapezoidal, with the triangle's tip cut off. In profile, the back edge of the stem coincides with the bearding line of the traditional rabbet. From the inside of the boat, only the flat aftersurface of the stem will be exposed. The acute pockets found in a traditional stem, between stem and planking aft of the bearding line, will be happily missing.

The forward edge of the stem is composed of two curves. Where the hull is sharp and the stem comes to a point, the forward edge is the bow profile, to inside of plank. Where the angle is blunter, the curve is cut back to coincide with the curve of the bow profile to outside of plank, offset by the thickness of the stem shoe.

The stem's width, or sided dimension, is constant. It needs to be thick enough to provide sufficient surface for the plank landing, but not so thick as to make the stem overly bulky. The stem can be built up by stacking stock to the thickness desired using scarfed joints staggered through the stack. Then, the fore and aft curves are laid out and cut. The flats of the forward edge, if there are any, are marked. Finally, the varying bevel is planed flat between the back edge and the front marks. The stem has much shape to it, but it is quite easy to make.

Keel/floor

The keel/floor structure is another opportunity for cold-molding. The traditional keel is a heavy balk of timber placed on the centerline at the bottom of the boat. Its job is to provide a landing for the garboard (lowermost) plank and provide support for the mast step and ballast. Cross timbers, called **floors**, are placed upon it. Their job is to tie the two sides of the boat together and to distribute the load from the ballast keel bolts into the hull shell via the frames. The floors tie pairs of frames, port and starboard, together with bolted connections. Keel bolts run vertically through the floors into the ballast. The top of the keel along with its floors and lower frame ends forms a complex, hard-to-clean, bilge into which all loose trash falls.

The traditional keel floor structure has two fundamental weaknesses. First, the keel being

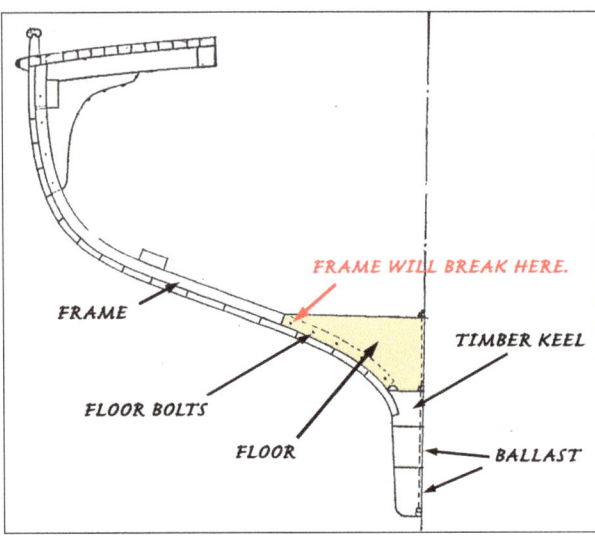

10-6 Traditional floors. A drawing from Chapelle's *Boatbuilding* showing traditional keel/floor structure. With a highly loaded modern rig, the floors often break the frames. When the frames break the mast structure is not supported by the hull. (Chapelle, *Yacht Designing*, p. 202.)

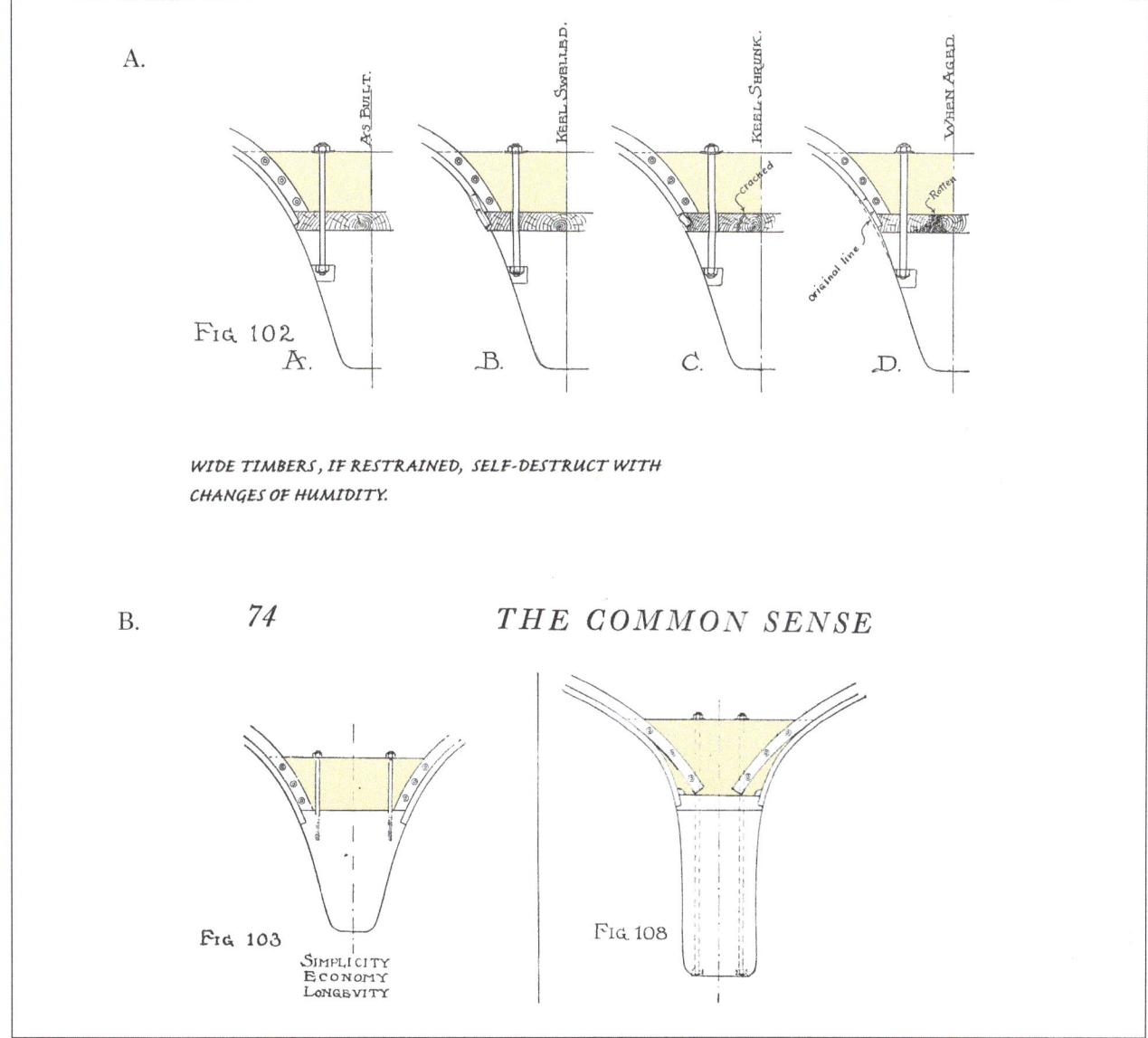

10-7 Francis Herreshoff's floors.

A. Francis Herreshoff's diagram shows what happens to a traditional keel when it swells and contracts due to moisture changes. Generally speaking, this is a problem with all wide timber, but with the wood keel it is particularly troublesome. The wood keel is securely restrained by the lead ballast, which does not move with moisture changes, and the result is a broken garboard and a rotten keel.

B. Herreshoff improved his keel structure by omitting the wood keel altogether! Surely, a radical idea. His figure 108 shows how to terminate frames on the floors to avoid housing them in mortices in the keel, avoiding pockets that collect water and cause rot. However, he does not show how to build the section where there is no ballast to act in place of the wood keel. We used Herreshoff's idea with great success with the *Alerion Class Sloop*.

wide and thick, it is subject to large movements as it dries and wets, causing problems with the attachment of the garboards and with rot. Francis Herreshoff illustrates the issue in his usual thorough way in *Common Sense of Yacht Design*.[1] His comments on keels are well taken about any large balk of timber. Wide timbers should be avoided where possible. His idea to use the ballast as a structural element was brilliant, and we used it to great advantage on the *Alerion Class Sloop*.

The second problem with traditional keel/floor structure is a geometric one—there is not enough space available to effectively bolt the floors and frames together. Space is particularly lacking at the mast step where the problem is most acute. What usually happens is that the frames are too small. They break at the floor-bolt holes. Then the loads of mast and ballast pull the garboard joints apart.

With cold-molded construction, we stabilize the keel timber by making it just thick enough to land the lower planks. We make it of stacked plywood. The plywood is stable athwartships because of its structure and it is stable vertically because it is thin. Because the keel will be highly stressed, we make it of dense, strong wood. Sapele, a very strong, durable species, is available in marine plywood, so it is a good choice. It is heavy, but down low, its weight is no disadvantage.

10-8 Floors for the cold-molded boat. In the cold-molded boat, the floors are replaced by a laminate of epoxy fiberglass that runs right around the bottom of the hull, tying the two sides together and providing tensile (hoop strength) capacity from the laminate on one side to the laminate of the other. This removes the the floors that cross and clutter the traditional bilge and makes for a clean bilge, the foundation of a happy ship.

The drawing shows the many lag screws that carry the ballast in place of the few bolts through the floors of the traditional boat. The philosophy of the cold-molded boat is to keep distributed loads distributed—don't concentrate them.

---

1. *Common Sense of Yacht Design*, pp. 73, 74.

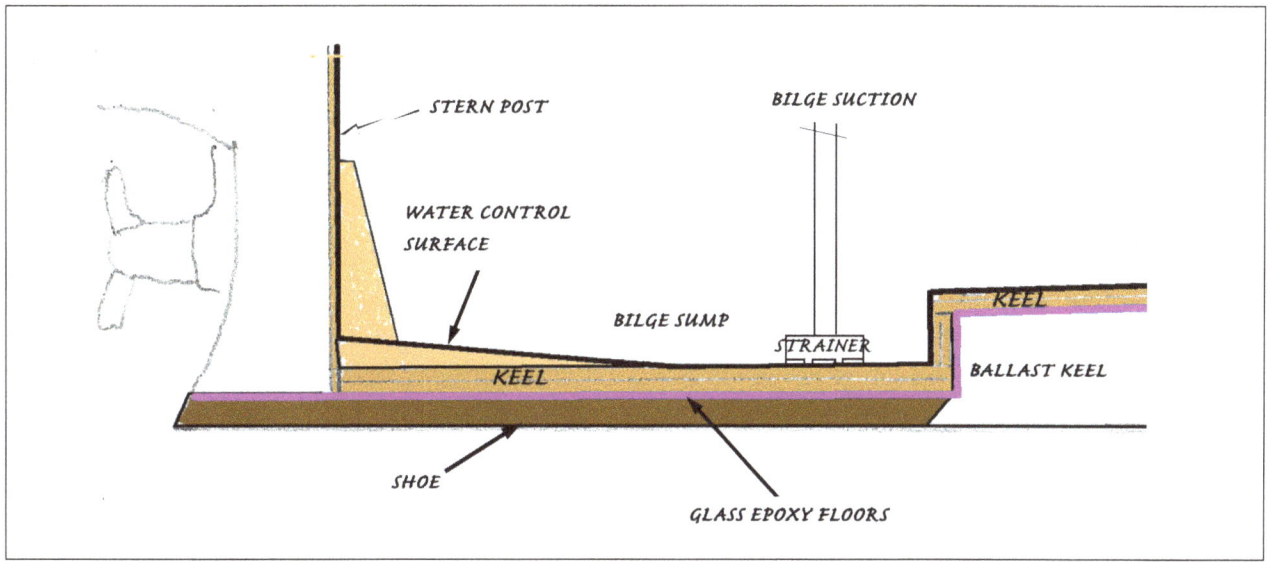

10-9 The sump. Every bilge needs a sump big enough to hold the bilge suction pump plus all the water that will flow back down the suction pipe when the pump stops. Especially in a shallow hull, the sump will be close to where the vessel will take the ground, so it should be strong and protected by a shoe of durable wood.

With cold-molding, the floors take a radically different form; in fact, as wooden members they disappear altogether. The conventional cross timbers are replaced by a swath of fiberglass that runs from within the hull skin on one side, under the keel into the skin on the other side, continuously along the entire length of the keel. This not only ties the two sides of the boat together but also ties the keel into the laminate. The laminate, like the frames of traditional construction, has transverse tension capacity and is where the tension load of the rig and ballast is ultimately headed.

The ballast keel is attached with metal fasteners so that it may be removed. With lead ballast, lag screws may be used, worked directly into the lead. The structural necessity is to transfer the load on the screw into the shell laminate. The fiberglass tape running under the keel and up into the laminate will do this as long as the geometry is not deformed. Placing the fasteners near the outer edge of the ballast close to the hull shell minimizes transverse bending loads on the keel. If the ballast is concentrated in a short area, we may wish to add reinforcing blocks that will stiffen the keel/hull shell corner. Remembering the density rule, these blocks should be of dense wood and might even have large G-10 (epoxy glass laminate) washers atop them with metal washers atop the G-10s.

Extreme fin keels and flat bottoms will require additional engineering for their specific conditions.

Deadwood

The deadwood is the fairing behind the ballast keel that streamlines it. Fin-keelers don't have it, but longer keel boats do, and while trivial in construction, it presents an important opportunity. All boats need a bilge sump; else a large swath of the bilge is cursed with perpetual wetness. The sump must be in the lowest recess and must be big enough to hold the volume of water that backflows from the discharge pipe when pumping stops. Figure 10-9 shows one way this can be done. It must be made strong, as being lowdown, it is subject to grounding loads. Being low down, there is no objection to using a very dense, durable wood like greenheart.

Horn timber

In many designs the horn timber is a fairly straight plank cut from a balk. If the rudder stock passes through it, the rudder packing-gland can be installed before it is put on the mold likewise, a mizzen mast step. With a canoe stern the horn timber will look more like the stem and can be treated in similar fashion.

Sole beam foundations

The cabin sole beam foundations are next attached to the sole mold (see figure 9-6). The foundations are wedge shape and double beveled to fit the inner surface of the hull shell. The loftsman will supply the angles. They are placed above the sole-mold raised up by a spacer whose thickness is the depth of the sole beam. The foundations may be glued either to a stringer, if there is one there, or directly to the planking as the hull is planked.

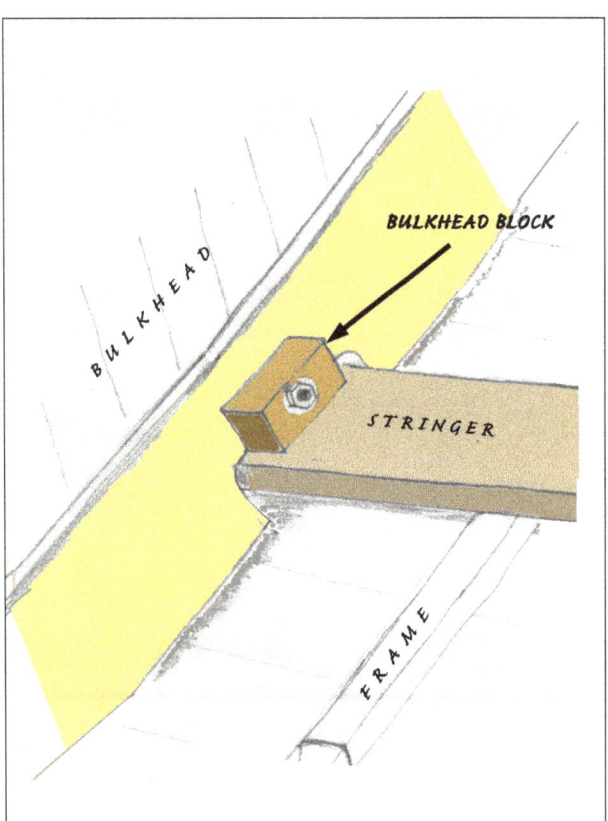

10-10 Bulkhead block. The bulkhead block is small and simple. It should be made of hard, dense, strong wood strongly and permanently attached to the stringer.

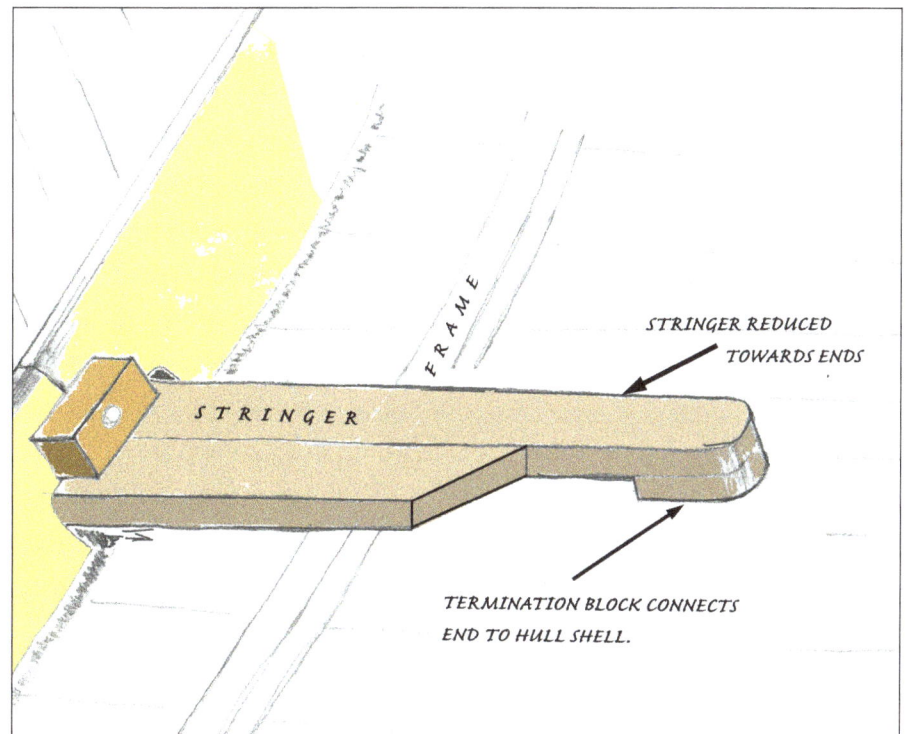

10-11 Stringers. If the stringer has much edge set, it may be made of two narrow pieces glued together after being bent in place. The second strip can be stopped before the end of the hull, saving some material. At the ends of the stringer add a block to fill the space between the stringer and hull. This may be done after the planking and laminate are complete to avoid making a hard spot. The termination block prevents snags on the stringer end and adds support to the shell.

Bulkhead blocks

With the perimeter of the hull shell defined by the beam shelf and the centerline structure, it is time to begin construction of the hull shell. The first step is to attach the bulkhead blocks to the bulkhead-molds. It is a simple process of bolting the blocks at the laser-marked positions using the laser-cut bolt holes. Bulkhead blocks are beveled because the stringers cross the bulkheads at an angle. Each one will be different. Its angle will be calculated by the loftsman and will usually be between 0° and 20°. The required angle will be marked on the mold and may be selected from a supply of blocks pre-cut with bevels of 0°, 2°, ... 20°. For a forty-footer, the blocks will be an inch wide, which means a 1° miss of bevel angle will give at most a $1/64$ inch misfit between block and stringer. Such is acceptable. The blocks are small and easily made.

Stringers

The stringers are scarfed together to full length. If the design calls for much edge set, they may be ripped in half lengthwise so that they will bend easily. The halves do not need to be glued back together, but a joint filled with glue is a joint that will not hold dirt or water, so it is not a bad idea. Then they are placed on the mold and fastened to the bulkhead blocks. Although the bulkhead blocks are removable from the molds, they are permanently attached to the stringers. The joint is important structurally; it needs to be

10-12 Bulkhead blocks on *STARRY NIGHT*. Usually the bulkhead blocks can be made to fall in lockers or other areas out of general line of sight.

CHAPTER 10 — BUILDING THE HULL

10-13 *STARRY NIGHT* framed out with planking begun. Her frames end on the lowest stringer. Below the frame ends, the bilge-molds take over the shaping of the shell. Note, *STARRY NIGHT* does not have an accommodation-type mold and there is no sole-mold. That idea came later.

carefully glued and fastened with large screws set in epoxy.

When made in two pieces, the upper piece may run full length and the lower half stopped 20% short of the ends to save a little weight and material. The ends of the stringers have short blocks attached to join them to the planking so that their ends do not make hooks that catch things.

The stringers are prominent in the finished boat. They can be pre-finished with varnish, before installation, but the back surface should be left unfinished so that no paint or varnish will interfere with a good stringer/frame glue joint.

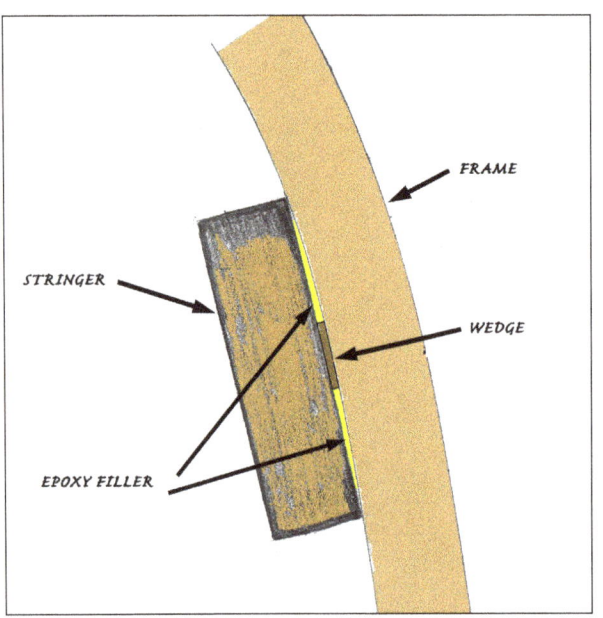

10-14 Tight frame curves. Where a frame is bent over a stringer with a strong curve, it is well to add a small wedge to prevent a hard spot.

Since the stringers project into the living space, their inboard two corners are rounded over.

When the hull is righted and the mold disassembled, the blocks remain in place to accept the accommodation bulkheads. The bolt holes remaining in the bulkhead blocks from their prior attachment to the molds will be reused for the actual bulkheads. The companion holes in the bulkheads may be drilled in place or, daringly, lofted onto the actual bulkhead patterns and drilled on the bench.

Frames

The frames are light and widely spaced. Their purpose is to space the stringers away from the hull plating, giving the hull structural thickness and further shaping the shell. The frames are square or flat, meaning their siding is equal to or greater than their molding. This makes them easy to bend, and it makes the vierendeel structure stable. Remember from figure 7-2 that the integrity of the viereendel is destroyed when the web element rolls. Being square or flat helps prevent this.

There is some skill required to bend frames, a skill which I do not possess; but Brad and Mike Pease do, as they demonstrated building *STARRY NIGHT* at their marine railway in Chatham, Massachusetts in 2009. The process, familiar to traditional builders, starts with the selection of the stock, minding species, grain structure and seasoning. For *STARRY NIGHT* the Pease brothers used white oak sided 1½ inches, molded 1 inch. With simple equipment they heated the strips and bent them over the stringers, 30 frames to the side. It is an easy job if you know how to do it and the Peases were masters. The frames can be given a tight roundover on their inboard corners. Their outboard surface will be hard against the planking, so those corners should be left sharp.

The bottom of the frame will begin either on the lowest stringer or on a bilge-mold. It will be bent across the other stringers and end in a notch on the beam shelf. The frames touch nothing but the stringers and the planking, so they may take their own path between sheer and heel. They will become more canted towards the ends of the vessel. Being allowed to cant, they do not require twisting. Being parallel to both plank and stringer, they need no bevel. They are much easier to make and install than traditional frames.

The frames are screw fastened and glued to the stringers. In areas of tight curvature, shims may be placed under the frame to prevent the fastener from pulling it down to form a hard spot. The shims may be ¹⁄₃₂ and ¹⁄₁₆ inch thick, and the loftsman can calculate which crossings will want them. Thickened epoxy should be used to fill the resulting pockets.

After the frames are all in place, the back edges of the stringers may be rounded over.

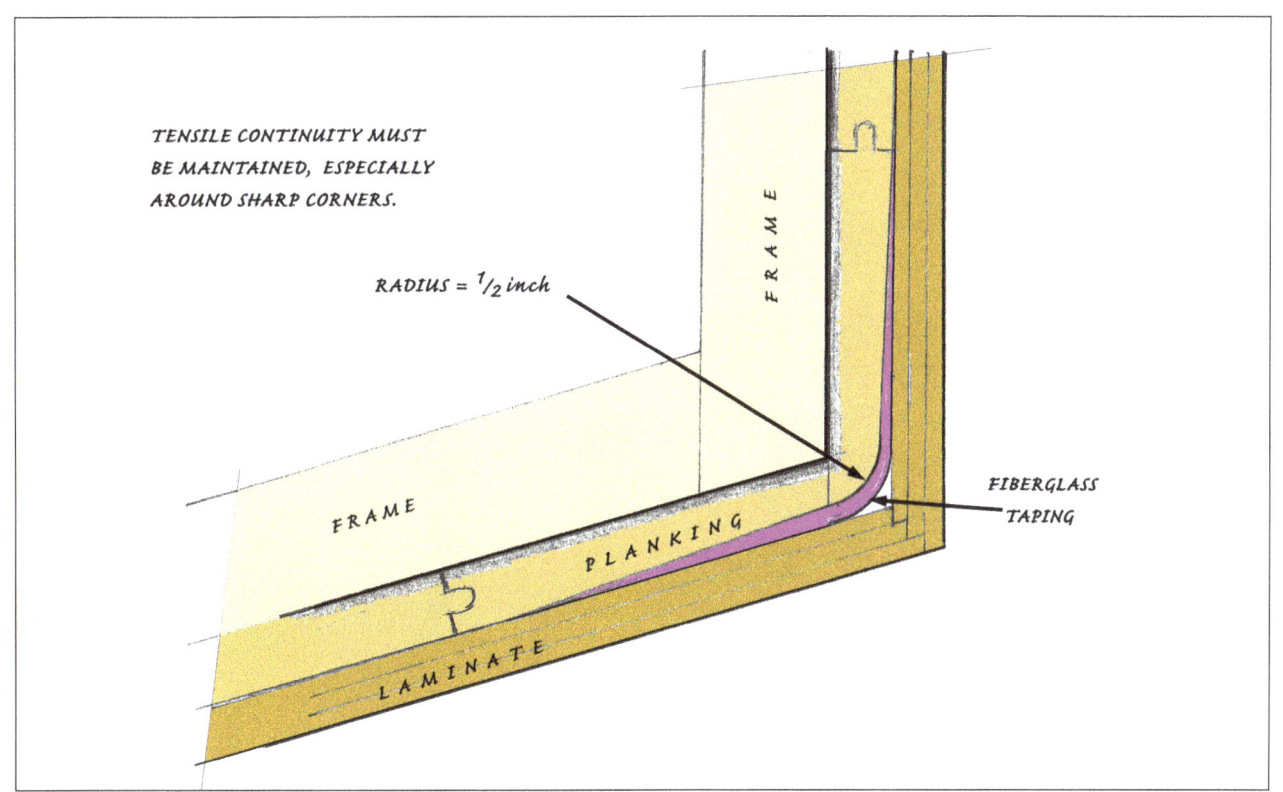

10-15 Tensile continuity (hoop strength) at sharp corners. Versatile as cold-molding is, wood veneers will not bend over a sharp corner. Around such corners, fiberglass tape is set in epoxy over the planking and under the laminate. The sharp corner is rounded over to a ½ inch radius to accommodate the glass tape. The laminate will cover the tape and build the corner back to sharp if desired.

Then the frames and backs of the stringers may be finished with a preservative.

Hull plating

Plating is an engineering term for the skin of a hull, the stuff that keeps the water out. The term likely derives from 19th century Scottish iron shipbuilding on the Clyde. On our boat, plating will consist of two parts. First, planking will be attached over the frames running fore and aft. Then, over the planking, a laminate of two or more layers of diagonal veneers will be cold-molded.

First the planking: a forty-footer may have planks ⅝ inch thick. We wish to make things easy, so the plank will be narrow enough, about 4 times its thickness, to edge-set without buckling. Edge setting eliminates the need for spiling, which is elegant but laborious. To keep plank material grain irregularities from producing bumps, the planks will be tongue and grooved, so adjacent planks will pull crooked ones back into the proper surface of the hull.

Planks are edge glued and glued to the frames. They are fastened lightly to the frames to hold them while the glue sets. The fasteners may be screws or, perhaps, boat nails driven by nail gun.

Should the planks be tapered? Good question. The joints between the planks will be visible, though not prominently, inside the finished hull. Non-tapered planks will rise toward the ends in the upper half of the hull and droop in the lower. By starting the planking in the middle and working both upwards and downwards, the rise and fall will be equalized, reducing the amount of rise and droop. This effect can be eliminated by tapering the planks. Whether it is worth the effort involved to make the pattern more pleasing is a question the builder must decide for each boat.

With computerized lofting it is possible to spile the planks and lay up the plating as we did with the *Alerion Class Sloop*. In 2012, when Pease Boat Works began building the *Alerion Class Sloops*, we replaced the traditionally generated plank shapes with computer generated ones. Laurie McGowan did the job with a Swedish CAD program. This was a new technique at the time, but we got a useable result and Laurie has perfected the technique since. Using spiled planking requires more care with the layup and final fitting of the planks. The reward is that if spiled planking is used on the outer layer of the laminate, the hull may be finished bright, the ultimate in "yachty" appearance.

Laminate

A most beneficial novelty gained from cold-molding is the achievement of hoop strength (**RULE 5**), or tensile-carrying ability, in the plating of the hull skin. Multiple layers of thin planking are bent and glued around the

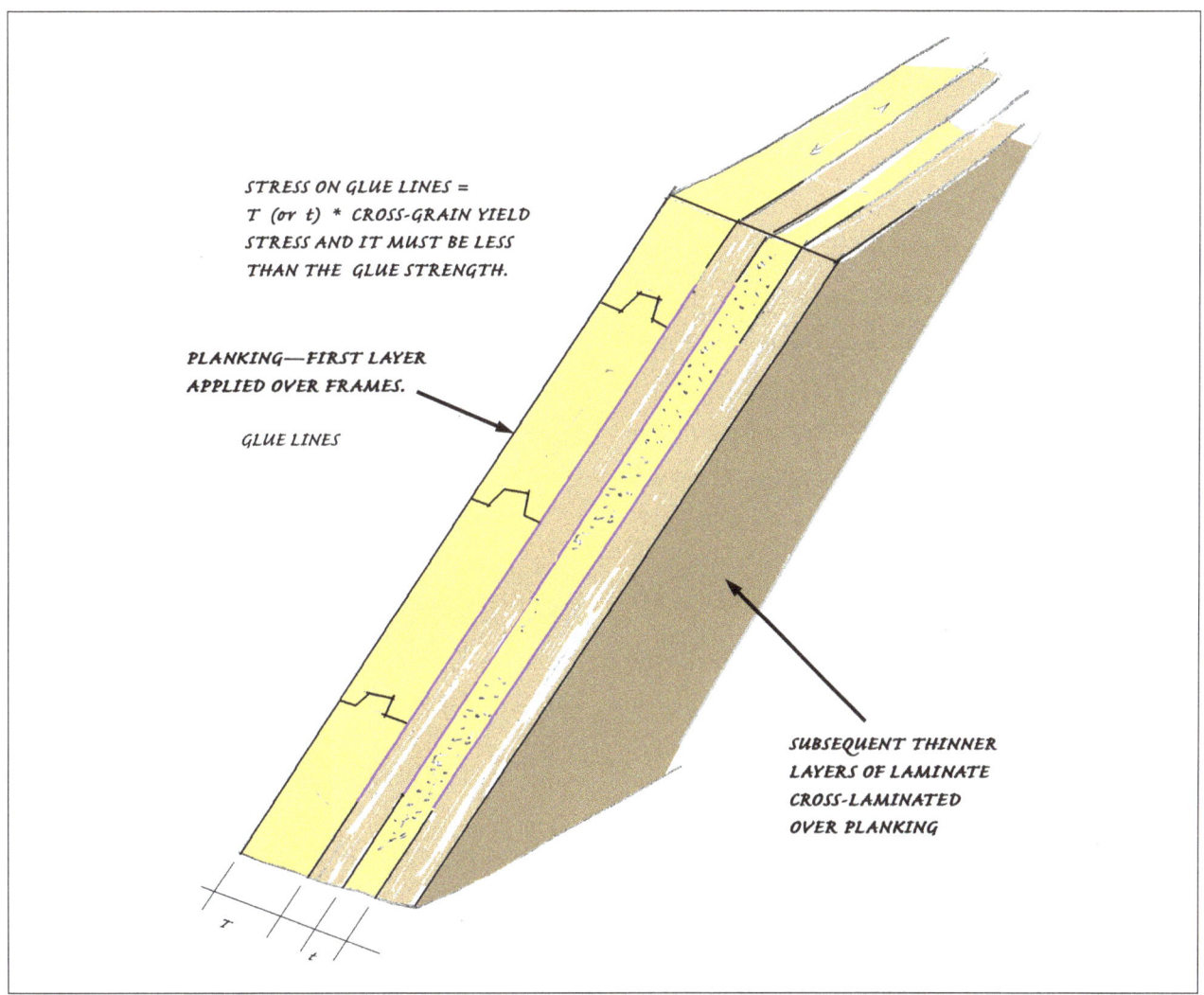

10-16 Hull plating. This simple scheme of plating is much used today. It consists of two parts. The planking runs fore & aft and is thick enough (T) to give a stable layer to laminate over. And the diagonal laminate is usually made of ⅛ inch thick (t) veneers. Boats built this way are painted to cover the unsightly diagonals and avoid the necessity of carefully fitting their joints.

curves of the cross sections, replacing the metal-fastened frames of traditional cross-banded planking. This not only eliminates the metal fastenings, it also eliminates the seams between the planks. The result is a one-piece structure with multi-directional strength.

The diagonal planking easily satisfies the hoop strength requirement, until one comes to a sharp corner. The thin planking will bend around strong curves, but it will not bend around sharp corners such as chines, the sheerline, or the stem. In traditional construction tensile continuity around these corners was provided by large pieces of wood—chine logs, shear clamps, beam shelves, and stems. Traditionally these are carefully shaped and bolted together at frequent intervals. They are stressed across their grain. They make up for their weakness across the grain by their bulk. Using epoxy bonding. we can substitute for these complex members fiberglass tape set in epoxy. The tape can be applied within the laminate so that it never sees the light of day and is protected by covering layers of wood. One-sixteenth inch of epoxy fiberglass is equal in tensile strength to three inches of cross grain white oak or six inches of Douglas fir.

Therefore, after the hull is planked, the next step is to epoxy-bond fiberglass tapes over all the sharp corners. To ease the tape around the corner, the corners are rounded over to a radius of ½ inch. On a forty-footer, 8 inches is wide enough for the tape. One or two layers will develop the required hoop strength. The rounded corner may be made sharp again with the subsequent layers of laminate. At this stage the beam shelf is adjoined to the hull planking with tape. This requires gluing the tape to the underside of the horizontal surface of the shelf. The vacuum bag will be helpful.

Once the epoxy taping is done, the required number of layers of veneer are laminated over the planked hull shell. Some designs will call for extra layers in the bow and garboard areas. The extra layers should be put on first so that

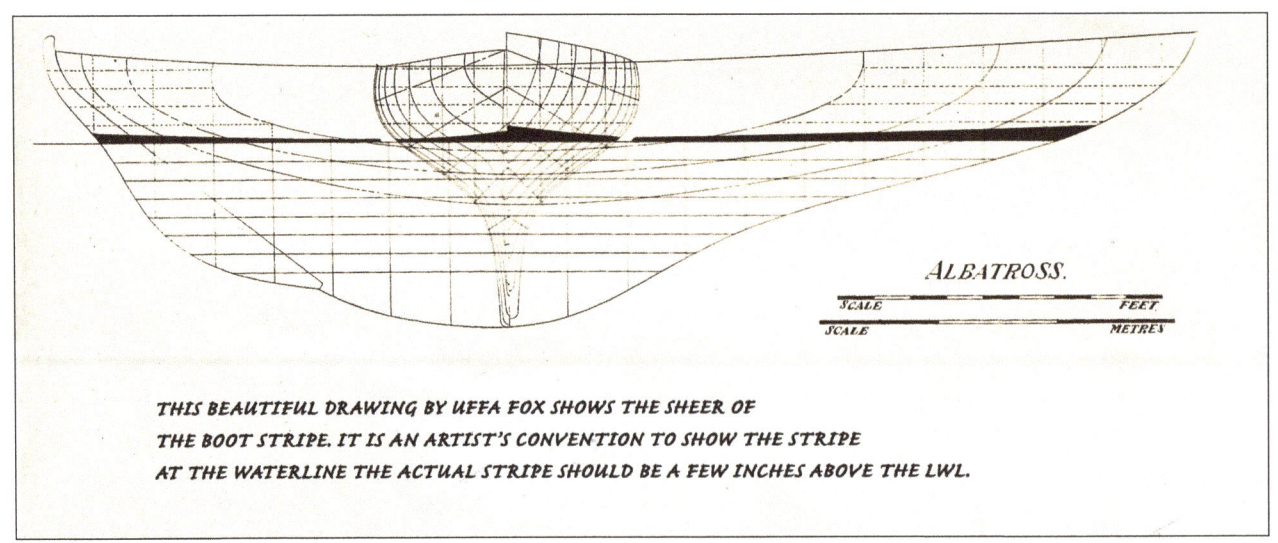

THIS BEAUTIFUL DRAWING BY UFFA FOX SHOWS THE SHEER OF THE BOOT STRIPE. IT IS AN ARTIST'S CONVENTION TO SHOW THE STRIPE AT THE WATERLINE THE ACTUAL STRIPE SHOULD BE A FEW INCHES ABOVE THE LWL.

10-17 **Striking the waterline.** The last job with the hull upside down on the mold is painting and that includes striking the waterline, or the boot stripe. The bottom of the stripe is a few inches above the load waterline (LWL) and is straight. The upper edge is a gentle sweep, a little higher in the bow than at the stern. It can be generated from a three-control-point cubic spline. Even a few inches above the LWL, the stripe is subject to fouling, so it is best painted with anti-fouling paint.

the last layer covers the entire hull. This will save fairing time and make the laminate more water resistant.

Each year the laminate process becomes more sophisticated. In the 1970's building the *Alerion Class Sloop*, we stapled each layer, glueing as we went. After the glue had set, we pulled the staples and sanded off the excess glue before starting the next layer. It was painstaking, and we often left unwanted voids. Today most shops use vacuum bagging to squeeze the laminate against the planking under the high pressure of our atmosphere (about 2,000 pounds per square foot) while the glue sets. They will use nylon staples that are allowed to remain in the finished laminate. An even newer technique, called vacuum infusion, assembles all the layers, lightly held in place, and then, in one step, pulls epoxy into the entire laminate with a vacuum. This process minimizes the amount of glue used. It is faster and cleaner than doing the glue-up one layer at a time.

Painting

Once the laminate is finished, the hull is sanded and faired. It will be quite fair to start with, so the process is mostly making a smooth surface ready for finishing. This is the time to strike the waterline, known nautically as the **boottop**. The boottop will be marked, its bottom edge a straight line about 1% of the waterline length above the designed line of flotation. The top edge of the boottop should have sheer, rising in a graceful curve, from a low point about 60% of the way aft. It should rise a bit more in the bow than in the stern. The two lines are generally inscribed so that they will remain for many a future paint job.

We are done with the mold. We raise the work off the floor and disassemble the mold while it is still upside down. The mold parts go to the cabinet shop to be used as patterns, unless the cabinet patterns have already been computer duplicated. In that case they go to storage to await the next boat. Then with the boat hanging inverted, the interior surface of the planking is cleaned. Regardless of the care taken with the planking, some glue is likely to have seeped inside and now is the time to remove it. With the boat upside down the job is easier; the workmen are standing on a flat floor, and the detritus falls to the floor. Once cleared of excess glue, the interior planking surface is treated with preservative sealer.

The hull is uprighted. It is traditional practice to set the hull up with its waterplane level. While this is not necessary with our method of locating the interior, it may help prevent mistakes as our project moves on to the next stage—the installation of her interior machinery and accommodation.

CHAPTER 11

Installing the Interior

OUR PRIOR PREPARATIONS make this a short chapter. We have built the hull, cleaned it up, and finished the inside surface. All the bulkhead blocks are in precisely the right places to accept the bulkheads. The sole beam foundations are in place ready for the sole to drop in. But before we clutter the inside, any large pieces of equipment need to be installed—engines, tanks, and such.

Heavy pieces can be placed to advantage on the stringers. The stringers will distribute their load into a considerable area of hull surface and support them above the hull surface, allowing any intruding water and debris to fall free into the bilge.

If the cockpit is mostly independent of the accommodation, it should now be put in place. Each boat will be different, but generally the cockpit can be pre-built and installed into place just under the deck, which it will support.

With the cockpit and machinery in place, rough plumbing and electrical can be accomplished. Items like through hulls, scuppers, and bilge pumps and their plumbing, batteries, and heavy wiring should be put in place while the boat remains open.

CHAPTER 11 — INSTALLING THE INTERIOR

11-1 Engine beds of *STARRY NIGHT*. After the cabin sole was built, allowing the wrights to walk around inside, and before the bulkheads got in the way, her engine was installed.

Replacing the molds with accommodation elements

The actual accommodation bulkheads are then bolted to the bulkhead blocks. The sequence is sole, bulkheads, panels, finish plumbing and electrical, surfaces, and final painting. The normally expensive process of fitting the accommodation into the hull becomes simply a matter of bolting prefabricated pieces into predetermined sockets. Using the same attachment, they take the exact location previously occupied by the bulkhead-molds. No cutting, no fitting, no measuring, no particular precision is required. If the molds were imprecise, the accommodation pieces will be imprecise in just the same way. Even if there were a gross mistake, all will still fit together; in the end, the only difference a big error may make is an imperceptible change in the shape of the hull.

Building the accommodation pieces

While we are on the subject of the interior accommodation, let us digress from assembly of the boat for a moment and use the remainder of the chapter to discuss the prefabrication of interior pieces. The system we used for STARRY NIGHT's interior was easy and is compatible with the accommodation mold technique. It created a light, strong, and beautiful interior.

Although there will be a few three-dimensional pieces, such as the companion ladder, drawers, and saloon table, for the most part, the interior of an oceangoing sailboat can be constructed by assembling planar elements into bunks, benches, counters, lockers, and cupboards. STARRY NIGHT's interior consisted of 52 planar elements—19 bulkheads, 16 panels, and 17 surfaces.

Wood design must always deal with the linear nature of wood. Any flatwork will move greatly with changes of moisture content across the grain. A 4 foot wide piece, such as a bulkhead, might change in width as much as 3 inches as it goes from wet to dry. This is not acceptable. As discussed in chapter 5, plywood is stable under moisture change, due to its structure, so we used okoume plywood as the base for STARRY NIGHT's accommodation pieces.

STARRY NIGHT's horizontal surfaces were conventional and made of 12mm okoume plywood. When used as countertops, a ¼ inch thick veneer was laminated to the plywood to form a mahogany working surface.

Her bulkheads and panels are much more interesting. They were built up as pseudo framed flat panels over a base of 6mm okoume plywood. First, at the mill, the 6mm plywood was given an additional surface veneer of butternut. We chose butternut, a favorite for sailboat cabins, because of its beautiful color and its light weight. This brought the thickness to a little over ¼ inch. The plywood with butternut weighed about ½ pound per square foot.

If you have worked with ¼ inch plywood, you know it is flimsy and will not stay flat. If you remember chapter 6, you will know that the way to stiffen something is to make it thicker. We did this by glueing solid butternut framing, ½ thick, to one or both sides. The frame widths varied from 1½ to 3½ inches. The framing increased the effective thickness of the plywood to ¾ inch or, where it was used both sides, to 1¼ inch. The framing gave the thin plywood the appearance of flat paneling, which is a very pleasant look that has been used domestically for centuries. The framing stiffened the pieces sufficiently that a heavy man could lurch against a panel without sensing its deflection. The completed framed pieces weighed about 1.3 pounds per square foot, which compares well with foam core glass/epoxy panels.

A feature of the plywood-based faux paneling is that the paneling patterns on one side

CHAPTER 11 — INSTALLING THE INTERIOR

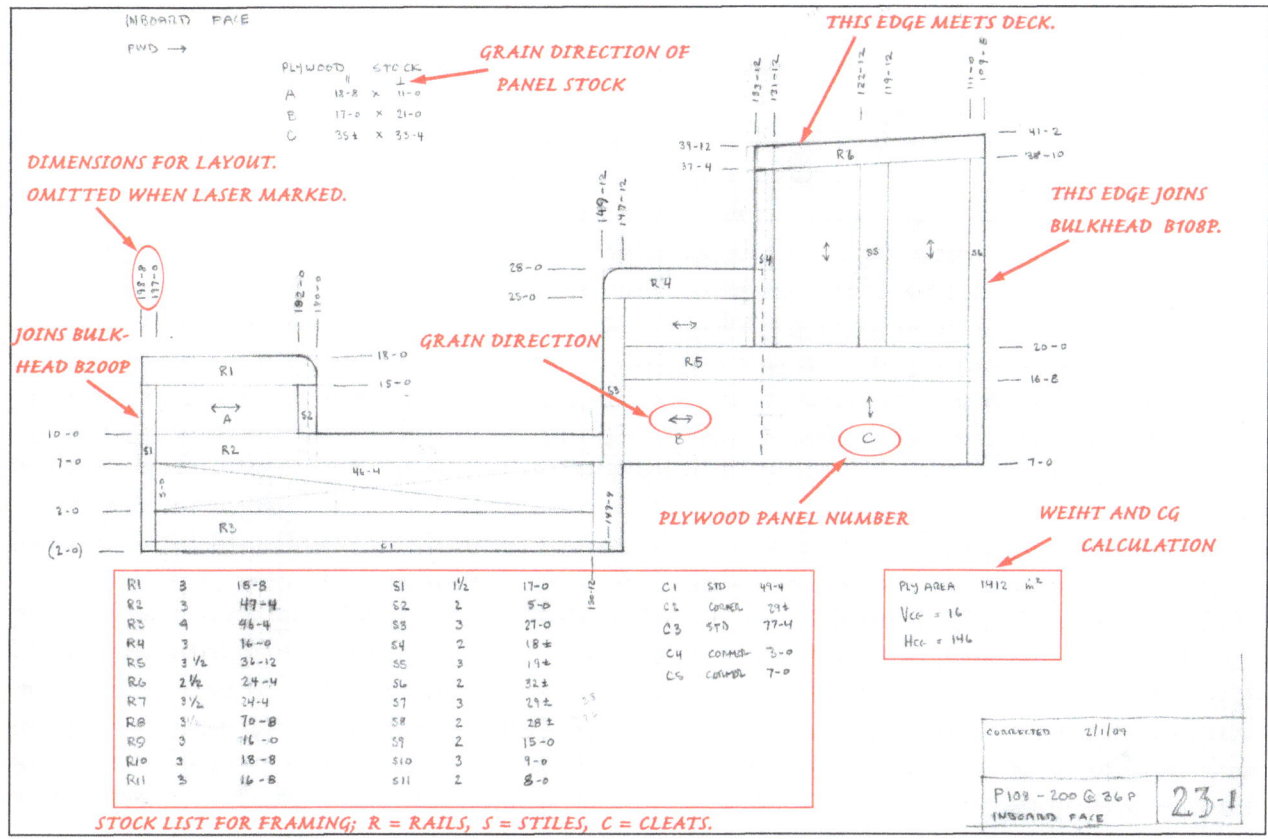

11-2 Bulkhead and panel drawings. The interior elements were made in the cabinet shop from shop drawings that spelled out every detail the cabinet maker needed to know. The drawings were 11 × 17 sheets drawn freehand. The whole job required about 70 of them.

of a piece could be independent of the pattern on the other side. A partition separating the aft cabin from the head used a paneling pattern appropriate to the head on one side and to the aft cabin on the other. Figure 11-3 shows the model I made to make sure patterns of the rails and stiles were consistent when the pieces were assembled into a cabin arrangement.

Figure 11-2 shows the shop drawing of the bunk face/galley counter back on the port side in the main saloon. There was a front and back drawing for each major piece. They were freehand drawn on 11 × 17 inch sheets and, as you can see, they include all the information the cabinetmaker needed to build them. Using these drawings, the cabinetmaker laid out each interior element on

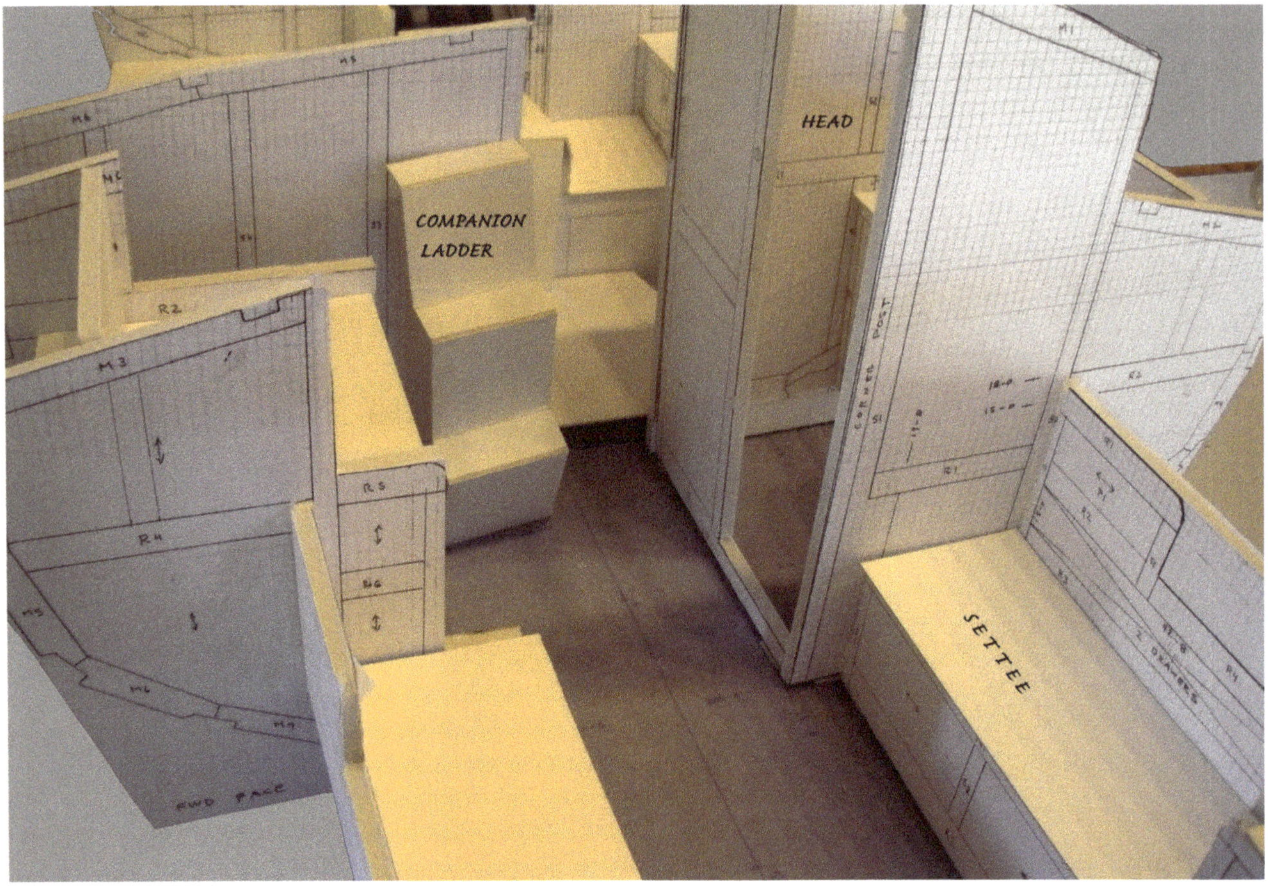

11-3 **Interior model.** I made a model by glueing shop drawings to foam core and cutting them out. It was used to visualize the joints and to confirm that the panels met the bulkheads with sensible patterns.

11-4 **Cabinet latches.** Wood of different species allows nice details like these latches, made of Mexican bocate, to give a boat character.

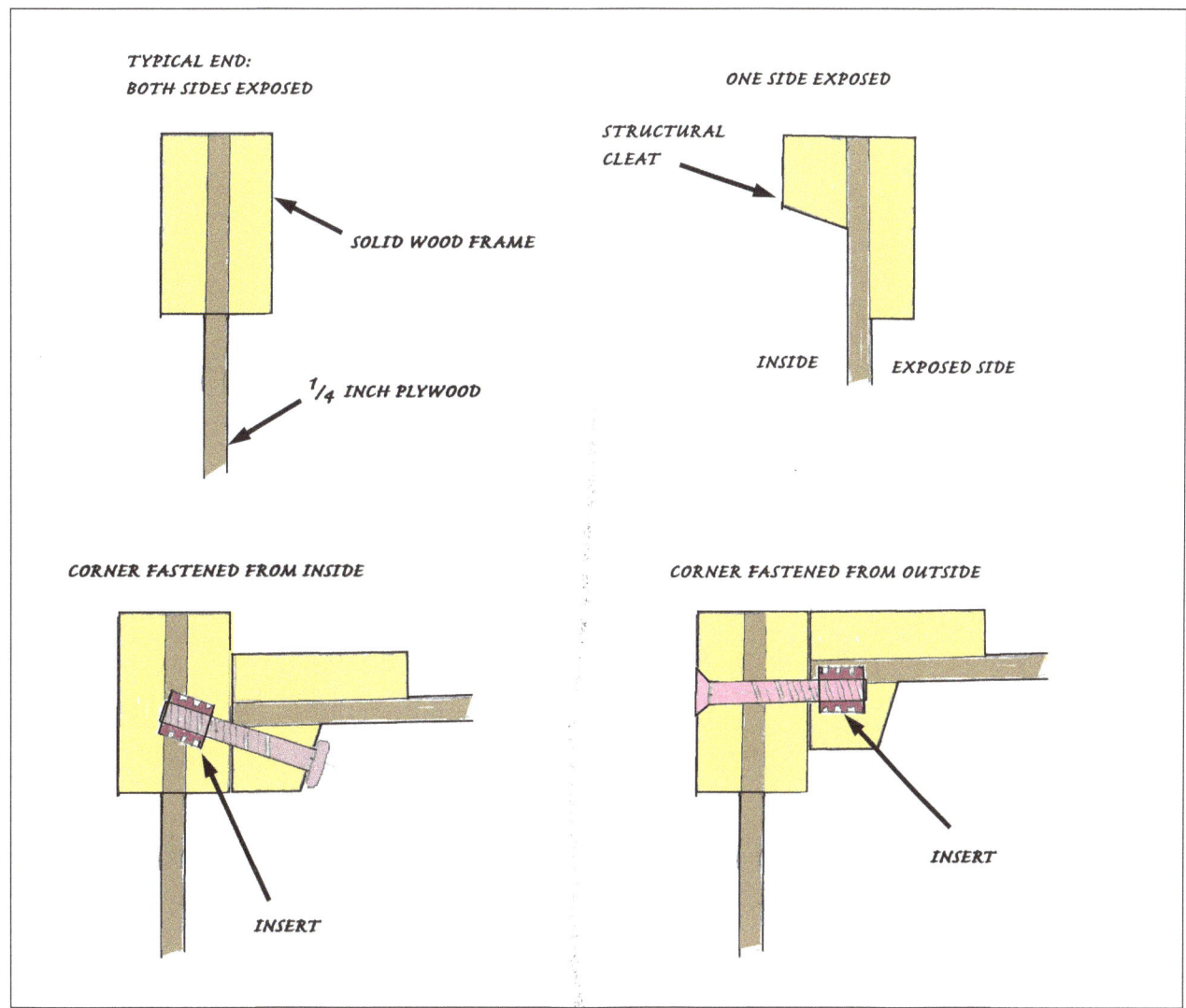

11-5 Panel bulkhead joint. The panels and bulkheads are joined together by machine screws, spaced about 8 inches vertically. They can be entirely hidden, but there are places where exposed screws will be more practical. The joints are made on the bench, so they will be easy to assemble and disassemble aboard the boat.

the butternut plywood and glued the framing to it. After gluing, the interior panels were given a 1/8 inch roundover on their frame edges. At this point cupboard and hanging locker doors were installed with their hinges and latches. Then the exterior edges were cut to exact size and the finished pieces went to the paint shop for varnishing.

Today, using the computer-generated loftings for the mold elements, the butternut plywood can be cut out and at the same time marked with the

11-6 The wooden cabin. This rich interior is built mostly of planar elements, the cabin table and companion ladder being the exceptions.

layout of the framing by the computer controlled laser cutter. Basically, figure 11-2 may be applied directly to the plywood, eliminating layout time and errors in the cabinet shop.

There remains the question of how to fasten the panels, bulkheads, and surfaces together. On *STARRY NIGHT* we used wood screws run in from behind the exposed surfaces. This gives a neat appearance with no fastenings visible, but having used the system more, I think a better way is to join the pieces together, on the bench, with machine screws running into threaded inserts. This gives a strong joint, properly positioned, that can be taken apart and reassembled many times. The inserts should be stainless, because they can be glued into the wood. The screws may be stainless or bronze: ¼-20 machine screws are a good size.

The end result gives a beautiful interior that everyone desires but few expect (**RULE 2**). Our interior is complete. Now we may close in the boat with the deck.

CHAPTER **12**

Building the Deck

WHEN SAILING, THE SAILOR works his boat from the deck. Off watch, he shelters under it. The deck, with its hatches and vents, acts as a membrane that allows passage of the good—sailors and air—and prevents passage of the bad—cold and wet. As discussed in chapter 4, the deck is a major structural element of the boat, it being the top chord of the beam that carries the primary loads. It requires strength, water tightness, utility, and beauty.

A lot happens on the deck. It provides the foundations for most of the boat's sailing gear, including sheet winches, genoa tracks, jib leads, travelers, bulwarks, chocks, lifeline stanchions, stowage boxes, windlass, anchor stowage, mooring cleats, spinnaker pole chocks, house and cockpit coamings, companionway hatches, ventilation hatches, dorade boxes, mast partners, life raft, and dinghy stowage. There is little activity on a sailboat that does not involve the deck. All this equipment requires careful design so that it can be easily worked and easily serviceable.

Structurally a deck differs from a hull because, being almost flat, it gets little strength as a shell. Rather, it is a planar surface spanning the hull. Due to its flatness it can be built from flat sheet

CHAPTER 12 — BUILDING THE DECKS

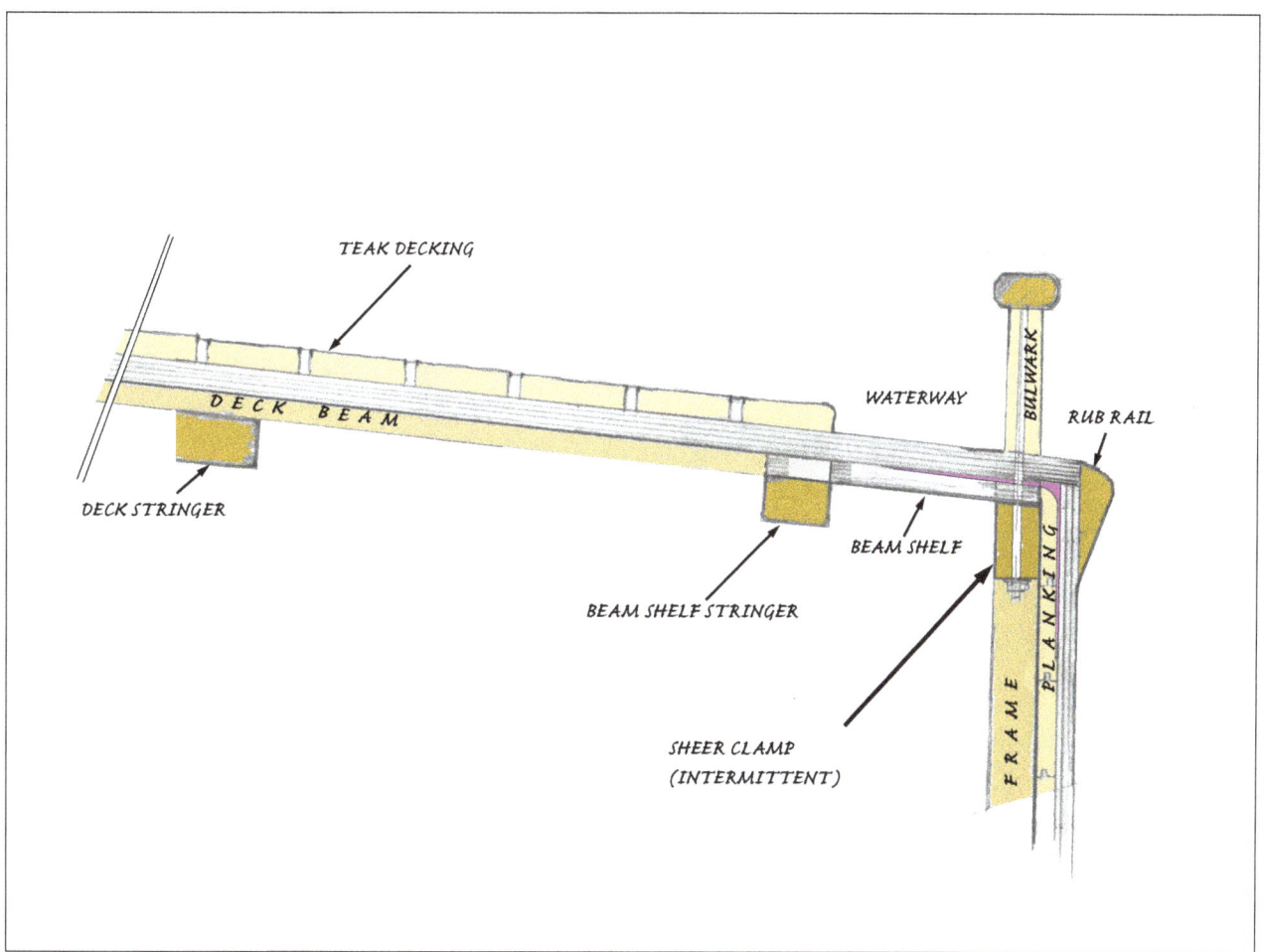

12-1 **Section of deck structure.** The deck is built mostly of plywood sheets and linear pieces of timber. Only the plywood pieces need shaping. Note the fiberglass taping that provides hoop strength around the sharp corner of the sheerline.

material that is tortured to take its small curvature. Carrying the primary sailing loads, the deck is stressed mostly fore & aft, so it wants to be longitudinally framed. It needs sufficient bending strength to be walked upon and to resist waves crashing upon it. Because of the need for hoop strength, it requires a certain amount of tensile strength in the athwartship direction, especially in way of the masts. And it must be locally reinforced to carry the loads of the deck gear.

In traditional construction, the deck is supported by large, closely spaced, transverse beams tied to the hull shell by the beam shelf and the sheer clamp. The beams are sawn or laminated to shape the crown of the deck and beveled to fit a longitudinal planking. They need to be big enough to span the width of the boat and are much bigger than required for hoop strength. The planks form the top chord of the hull beam and, once the minimum for walking is achieved, are so designed.

We will use a different deck to take advantage of epoxy bonding and computer-controlled cutting for building the deck on our cold-molded, stringer-generated boat. We will avoid difficult shaping and beveling of the beams and, of course, metal fastenings. The deck will be formed over longitudinal stringers that are attached to the accommodation bulkheads with the usual bulkhead blocks. The stringers will run straight, fore & aft, parallel to the centerline. Three or four per side are sufficient. Their ends terminate on the underside of the beam shelf. We will take heed of their location to maximize headroom in the passageways fore and aft.

Deck beams

Over the stringers we bend thin, flat deck beams. The deck beams do four jobs—they add to the hoop strength tying the two sides of the boat together, they stiffen the deck plating as it spans athwartships between stringers, they provide backing for the joints of the fore & aft plywood deck plating, and they provide extra thickness for the temporary screws used to lay the teak decking.

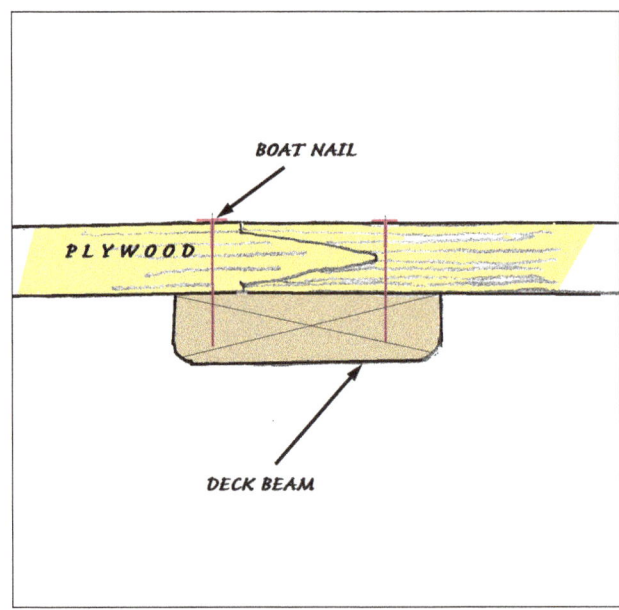

12-2 **Plywood deck joints.** The deck sheets are scarfed into 4 foot wide sheets of sufficient length to reach from side to side. Fore & aft, the transverse joints are tongue and groove jointed, with the joints landing on the deck beams, so that the entire deck becomes a single sheet.

The deck beams are the thickness of the beam shelf, thin enough to easily bend to the deck crown. For a forty-footer the beams might be molded ½ inch and sided 2 inches. The bottom edges are rounded over ⅛ inch. They land in notches cut on the beam shelf and are fastened to the shelf stringer that forms the bottom of the notch. They are spaced so that the transverse joints of the deck plating fall on them.

With the beams in place, the deck stringer's upper edges may be rounded over and their upper surface finished with preservative or varnish.

CHAPTER 12 — BUILDING THE DECKS

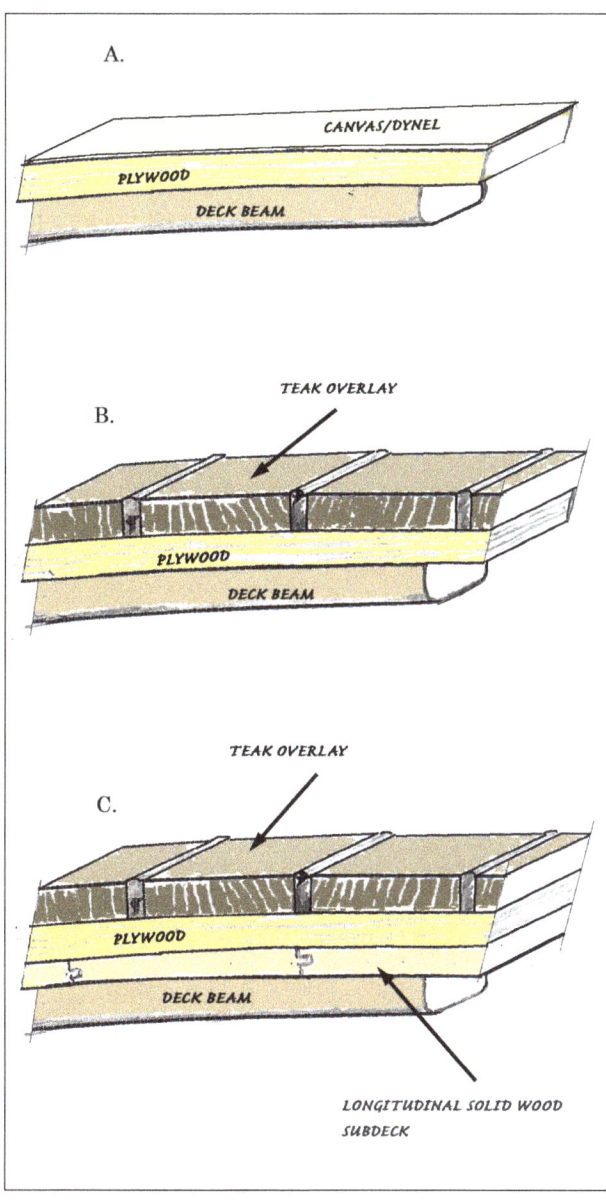

12-3 Three possible structures for deck plating. The simplest (A) is a layer of plywood laid across the deck beams. This can be finished with canvas and paint. A more modern technique uses dynel fabric set in epoxy. Many sailors will want a teak surface rather than the fabric (B). Finally, especially in larger boats that have a thicker deck, a layer of fore & aft planking can be put under the plywood (C). This gives a nice finish and gets most of the deck material running fore & aft.

Deck plating

There are several ways to structure the deck plating. The simplest uses a layer of plywood, finished with a layer of fabric for waterproofing (figure 12-3 A). We did this with the *Alerion Class Sloops* with good results. Later, owners wanted a fancier surface, so we replaced the fabric with a thin layer of teak, sprung with the sheer to look like a solid teak deck (figure 12-3 B). It cost more and was beautiful.

The plywood, adding transverse hoop strength to the light deck beams, is essential to the plating, as the beams are insufficient by themselves. The plywood sheet also forms an effective water barrier and also gives the deck diagonal strength to resist wracking. When topped with teak, the plywood can be made thinner; it will still have enough hoop strength. But the plywood sheets need to be joined both fore & aft and athwartships to make a structurally continuous sheet.

A third variation uses a thin layer of fore & aft solid wood planking under the plywood, three layers in all (figure 12-3 C). The planking is tongue and groove and runs straight, parallel to the centerline like the stringers. This provides a nice wood overhead down below and more of the deck material running fore & aft. Care must be taken that there is enough thickness of plywood to provide hoop strength, especially in the way of the mast, as the plywood and the small deck beams are all that hold the two sides of the boat together. The plywood may be glued to the planking and teak decking, which eliminates its need to be edge-joined fore & aft.

Building the deck

If you have the space, building the deck flat on the shop floor expedites joining the panels that make it up. Building it upside down allows all the reinforcement and hardware backing blocks to be attached. It allows the underside to be painted and finished even to the extent of adding lights and wiring. We built decks for the *Alerion Class Sloops* this way. But with larger vessels it is not so easy—the deck becomes large and unwieldy. A large sophisticated shop might have the room and equipment to turn over such a deck and lift it into place, but such a shop will be the exception rather than the rule.

A better way to build the deck is to assemble it piece by piece over the deck beams. A forty-foot boat will have as many as fifteen 4 × 8 foot sheets of plywood that need to be joined structurally. The sheets should be joined into 4 foot wide lengths sufficient to reach all the way across the boat. Then the 4 foot wide panels are laid across the boat and field-joined in the fore & aft direction with tongue and groove joints landing on the deck beams. When the plywood is glued to the beams, it becomes structurally integral with the beams, making a stiff, strong T-beam.

There is a lot of lofting and layout work on the plywood deck sheets, and their production is an excellent application for computer lofting and laser cutting and marking. The deck carries much hardware and has a number of openings. Most hardware will need backing blocks on the underside, and openings will need to be framed. Certain interior equipment, such as overhead lights, will want risers. On the upper surface, the exact layout of the teak margin pieces will be needed, and the locations of the deck beams will be required for gluing the teak decking down. The loftsman will lay out the locations and shapes of all these elements, as well as the curved edges of the sheets. His loftings will go to the subcontractor, who will deliver back plywood sheets cut and marked by the usual computer-controlled process.

Before the panels are installed over the deck beams, the backing and riser blocks and any hatch framing should be attached to the underside. Bolt holes for mounting equipment are drilled. They will be re-drilled after the teak decking is laid. Then the underside of the panels is finished. The lofting will have laid out the faying surfaces where the deck beams attach on the underside so that they can be masked to prevent any finish from interfering with the glue.

There is an important point to be made about the extent of the deck in the bow. Although with traditional construction the frame tops are covered by the covering board (the outer plank

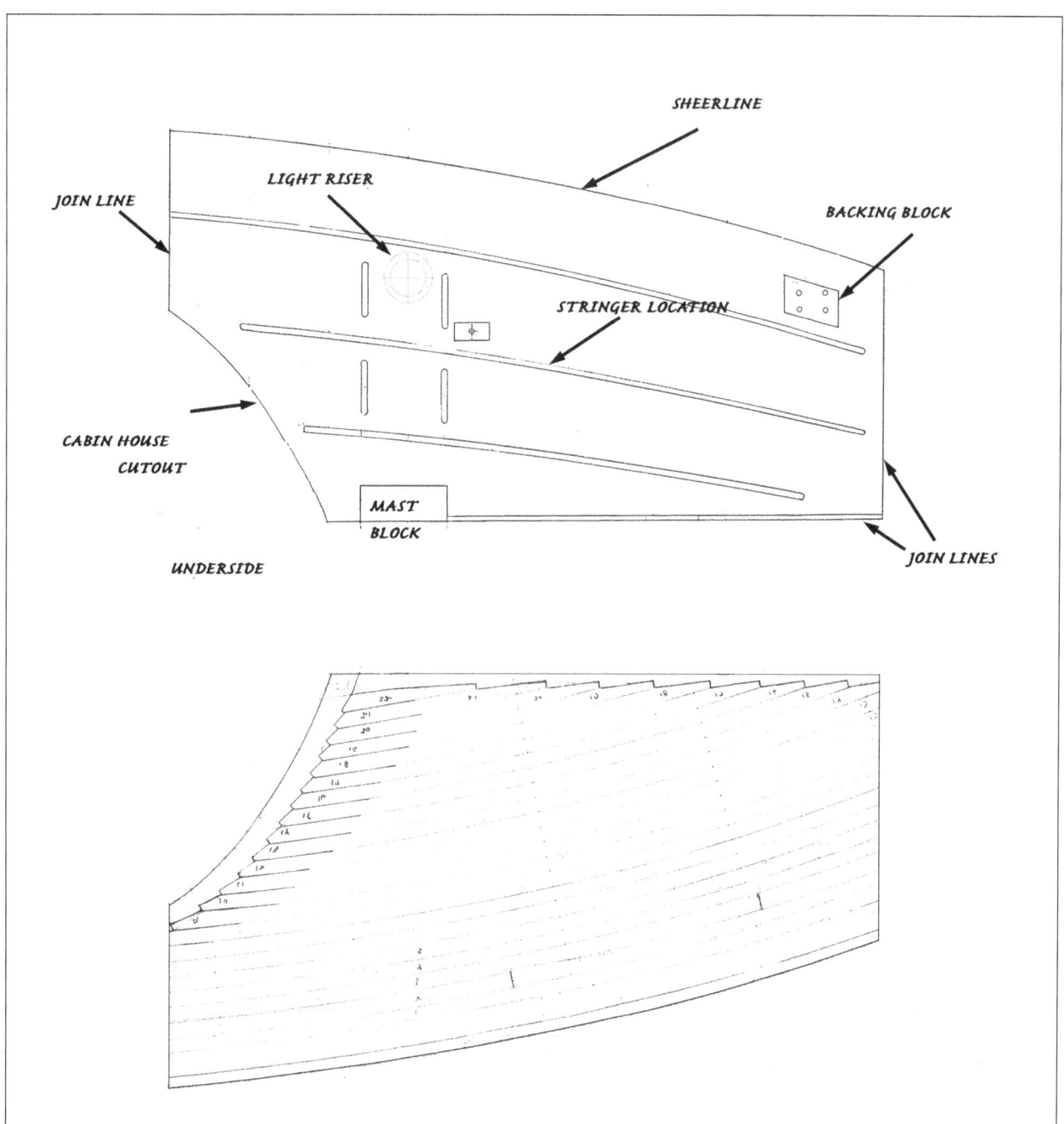

12-4 Deck layout. This is a shop drawing for a panel of the *Alerion Class Sloop*'s deck. Her deck was made up of six panels cut from five 4 × 8 foot sheets of Douglas fir marine plywood. With computer lofting and marking, the sheets would come from the cutting shop looking like this.

of the deck), it is common practice to bring the stem through the deck and use it to anchor the forward end of the bulwarks. This practice leaves a vertical joint through the deck surface.[1] Every moment the boat is in the out-of-doors, water, under force of gravity, will try to seep through this (and any vertical) joint. Eventually it will work its way through and run deep into the bow structure, subjecting it to decay. A cold-molded bow is much less vulnerable than the planked bow, but even so, there is no need for the stem to penetrate the deck. Never fight gravity! The deck plating should cover the top of the stem. Make the joint between hull and deck throughout the boat a horizontal one.

Teak decking

Laying a sprung teak surface over the plywood structural deck has become standard practice in recent years. The margin pieces, at the inner and outer edges of the decking, are sawn to shape. They may be screwed down temporarily while the glue sets. Then the screws should be withdrawn and their holes filled with plugs the full thickness of the teak. As the teak wears down, the plugs will not come out. A good glue

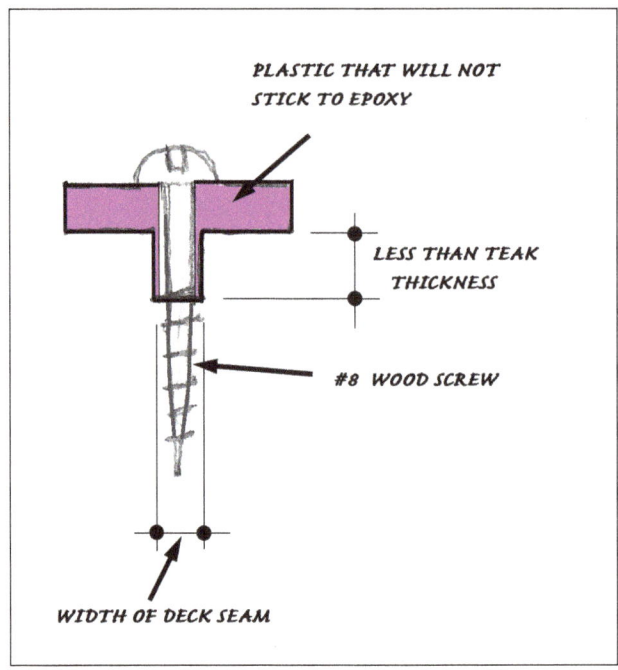

12-5 Hold-down jig. This simple spacer sets the calking groove width and holds down the inner edge of the teak plank as the glue dries. It is then removed. It must be made of a material that does not stick to the decking glue.

to use is MMM's 5200, which is a polyurethane. It has a lot of tenacity and a lot of flexibility—the teak will not come up.

After the margin pieces, the teak planks are glued into place and held temporarily with hold-down jigs, screwed to the subdeck through the gaps in the planking. Being strongly edge-set, the teak strips will try to lift their inner edge. If the hold-downs are placed to fall on the deck beams, there is plenty of thickness for the screw to get a good bite. It will be able to hold the teak down. When the glue has set, the screws and hold-downs will be removed. The screw holes

---

[1]. Halsey Herreshoff, commenting on our *Alerion Class Sloop* prototype, pointed out that all vertical joints will eventually leak. Regardless of the caulking, gravity is constantly trying to pull water in—all the time. Horizontal joints are equally vulnerable only when the boat is torrented by water or strongly heeled, and that is a small portion of the useful life of a boat, so horizontal joints are much less vulnerable.

will be filled by the seam compound that fills the joints between the planks. The resulting deck is pretty and will give a good foothold. It will be serviceable until it has been entirely worn through in thirty or more years. At that time, it can be restored using the same technique.

Houses and hatches

At sea, in extreme conditions, all deck openings are a threat to the watertight integrity of the boat. The easiest way to sink a boat is for a wave to sweep off her houses and downflood her. Worse than being swept is to be knocked down or rolled over and suffer the leeward house coaming broken (see figure 4-9). There are two forces that will break the coaming: the shearing loads at the deck joint and the bending loads midway up the coaming.

First, the sheering loads. The cold-molded boat treats the house deck joint as a sharp corner. But it is a little different from the chines because the corner between house and deck is an internal one. We use the same fiberglass tape to provide a lip that will resist shear. We will use a series of studs to provide tensile continuity (hoop strength) that the tape will not provide on an internal corner. The building process is to glue a margin piece at the edge of the opening, against which the glass tape can be glued to form a lip. The coaming is then fastened down on the outboard side of the lip. The teak decking covers the glass tape.

For the lip to work, the coaming needs to be strongly fastened vertically, holding it down to the deck. This is done by studs glued up into the coaming and bolted down through the deck beams. Stainless steel (as opposed to bronze, which does not glue well) studs embedded in epoxy will minimize the leakage and condensation issues. The Gougeon brothers give an analysis of glued studs and bolts in their book *On Boat Construction*,[2] a very good early book on cold-molding that discusses many of their inventions in the field.

Second, the bending loads—the downflooding disasters from knockdowns have mostly been reported by traditionally built wooden boats whose coamings are planks set on edge with grain running horizontally.[3] The coaming planks are not very thick and are weakened by large port openings. On knockdown, the sea's impact loads the cross grain of the plank, so it is not surprising that the coamings split along their length. This does not happen to the traditional hull because of the frames. It does not happen to the cold-molded hull because of the diagonals in the laminate. So the coamings need to take one of those forms also. Either we can make it of plywood with vertical strength laminated in, or we can use vertical frames to give cross-grain strength, taking care to strongly secure the ends of the frames to the deck and housetop.

2.. Gougeon Brothers, *On Boat Construction*, p. 301.
3. Coles, *Heavy Weather Sailing*; Smeeton, *Once Is Enough*.

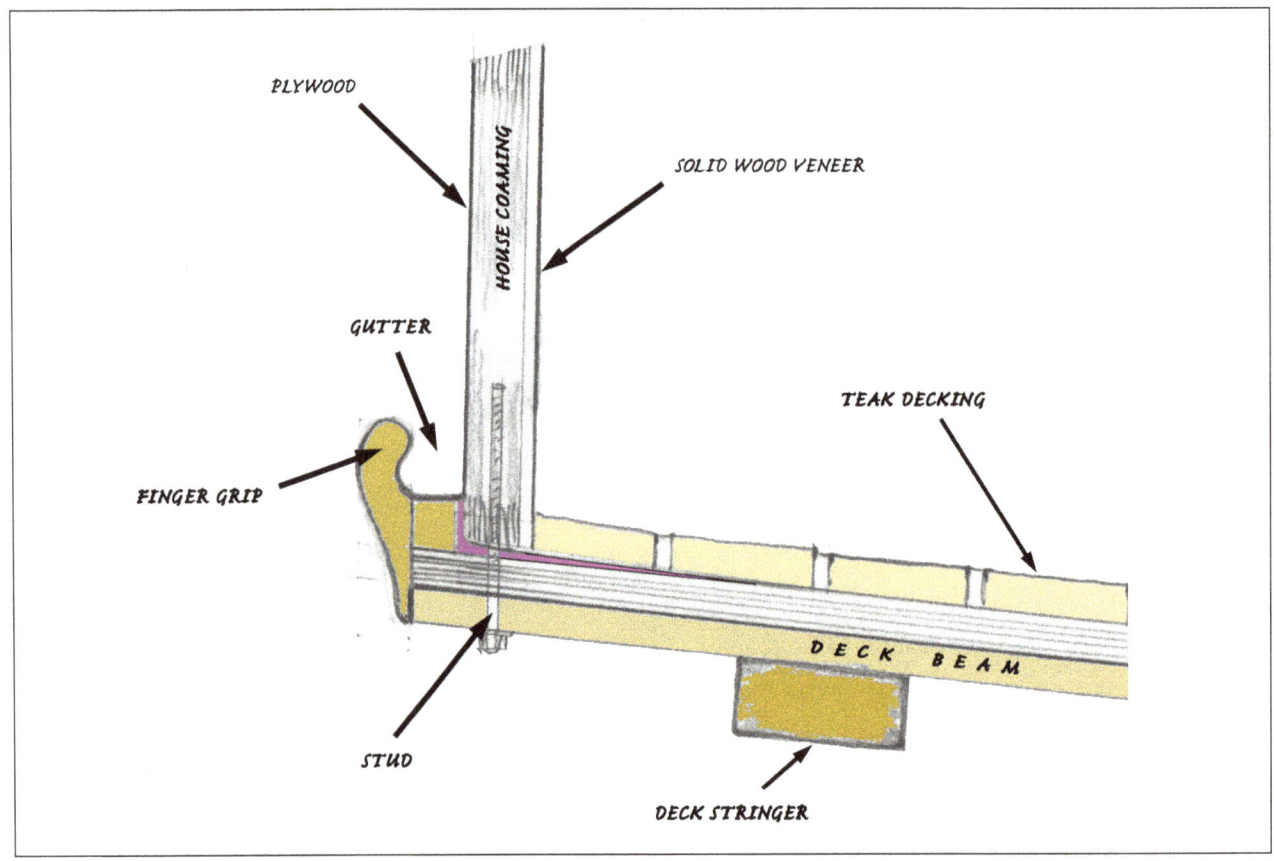

12-6 House rising from deck. See also figure 12-1. Notice the fiberglass tape running in the interior corner and the stud to hold the house down onto the deck. The gutter finger rail is a very nice detail.

A very neat finish for the inside edge of the deck is to cap it with the finger grip shown in figure 12-6.[4] This not only gives a very useful handhold, it also acts as a gutter to catch drips coming through the portholes. The bulkheads must be designed to allow the gutter to pass through them. Then, at the low point, there must be a drain to carry the collected water into the bilge.

With the deck built, the boat is structurally complete, but before ending, let's take a chapter to look at all the things that happen at the rail. This is where hull and deck are held together by the beam shelf. It needs some discussion in detail.

---

4. I discovered this on my Abeking and Rasmussen–built IMPALA.

CHAPTER 13

At the Rail

THE RAIL, THE CORNER where the deck joins the hull, is a busy place. Here, the bulwarks rise, the waterways run, the chainplates spring, hardware clutters, and the rub rail protects. Cold-molding and epoxy bonding make possible new configurations. The builder must work out the details carefully to keep the elements from interfering with each other.

Bulwarks

Bulwarks are a distinguishing feature of a blue water sailboat. Bulwarks provide the crew with a feeling of enclosure that cannot be felt on a deck with an unraised edge. They also provide positive foothold along the lee side of a heeled vessel. They are vestigial remains of ships' waist-high wooden rails. On a small yacht bulwarks need be only about 4 inches high to do their job.

The usual arrangement has the bulwarks built as a continuation of the topsides. This is a pleasing-looking arrangement, but to keep water out of the hull plating, we need the deck to cap it. So the question becomes, how are the bulwarks to be fastened through the deck and into the hull? The joint must be strong,

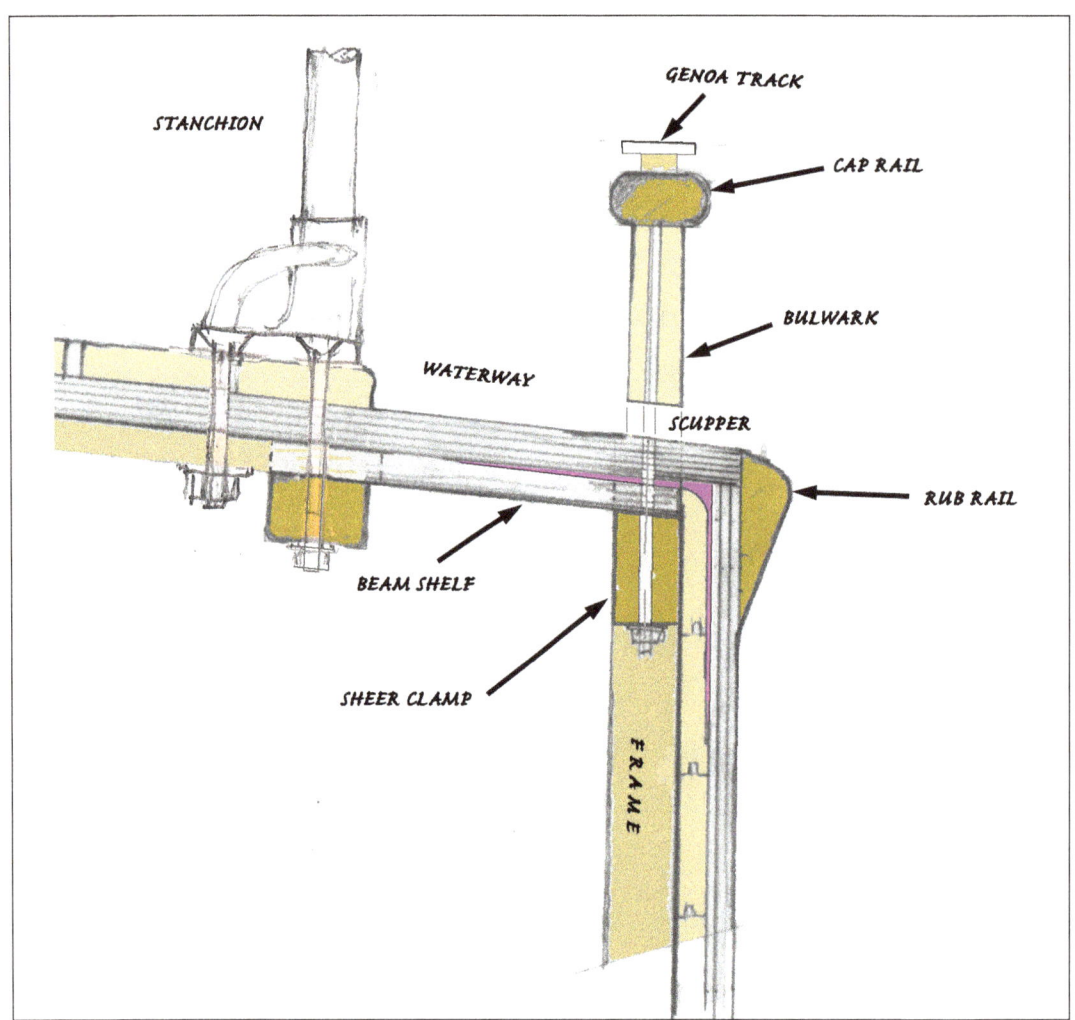

13-1 The rail is a busy place. This sketch shows stanchions inboard, the bulwarks inset the thickness of hull plating, the waterways clear, and the scuppers cut through the bulwark and directly overboard. The rub rail should be the most outboard element. It should be shaped so that it cannot hang up on a dock edge.

because bulwarks may carry a lot of sideways load. Traditionally it is made by using metal drift pins running through the covering board and into the sheer plank. It is a strong joint and looks well—the bulwark can be placed right to the outer edge, flush with the planking. The trouble is that when strained, the metal fastenings will waggle open their holes enough to draw water up into the bulwark and through the covering board into the sheer plank. This action caused the entire corner of my Abeking and Rasmussen–built *IMPALA* to rot. Her genoa tracks are

CHAPTER 13 · AT THE RAIL

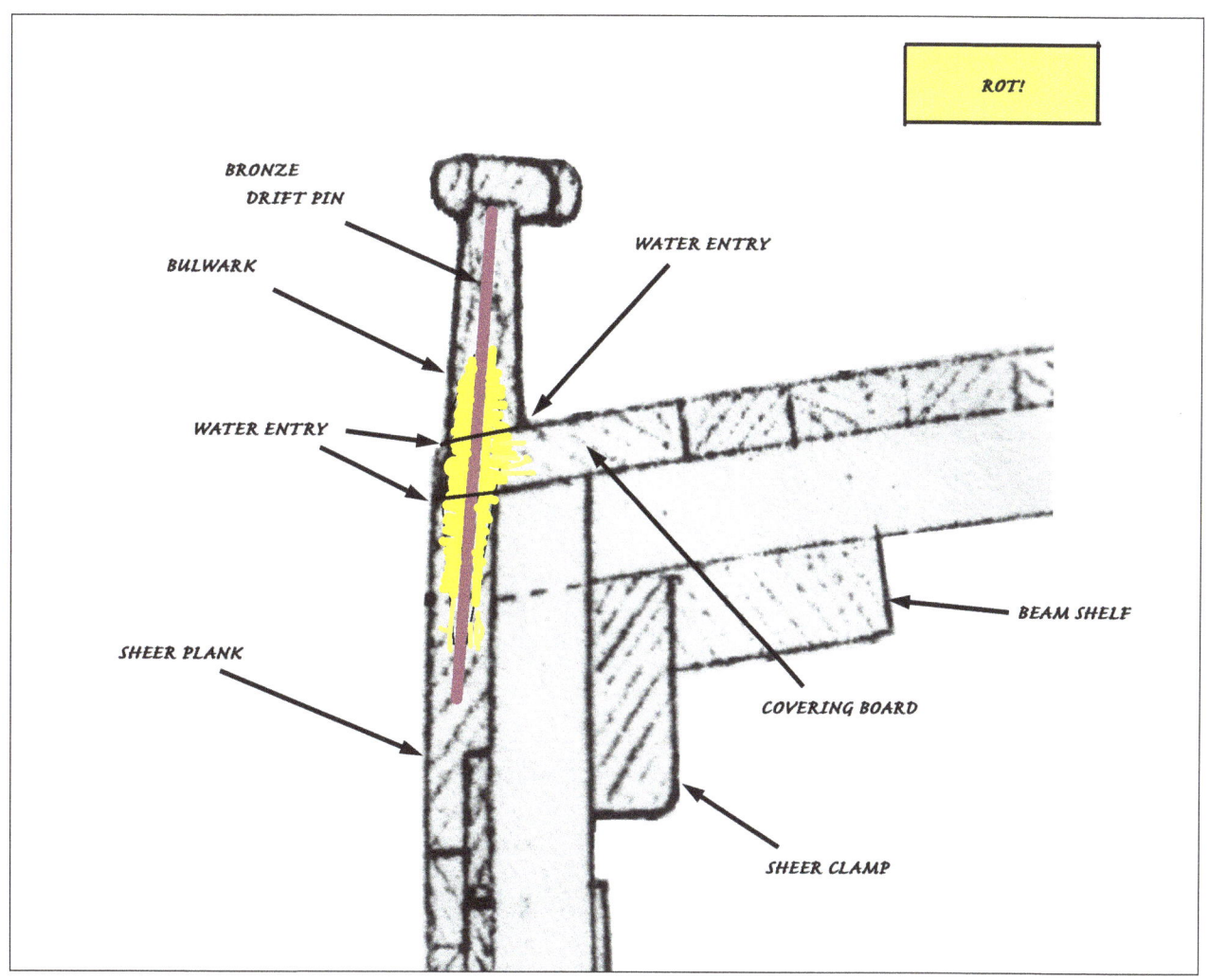

13-2 IMPALA's rail. As with many fine yachts, her rail is fastened through the covering board into the sheer plank with drift pins. As the water seeps into the joint, especially under strain from the genoa leads, the bulwark, covering board, and sheer plank all rot, creating a serious structural problem—and a big leak onto the two best sea bunks in the boat!

attached atop the bulwarks. The genoa sheets put large vertical and sideways loads onto the bulwarks, a pulsating load augmented by each passing wave. After 40 years water had entered and rotted her rail region (see figure 13-2). She was repaired reusing the original construction. Today, 25 years on, she is showing signs that she will have to be repaired again. Same cause, same construction, same result.

Epoxy bonding can improve the situation by omitting the metal fasteners and simply gluing the bulwark to the deck. But this makes repairs

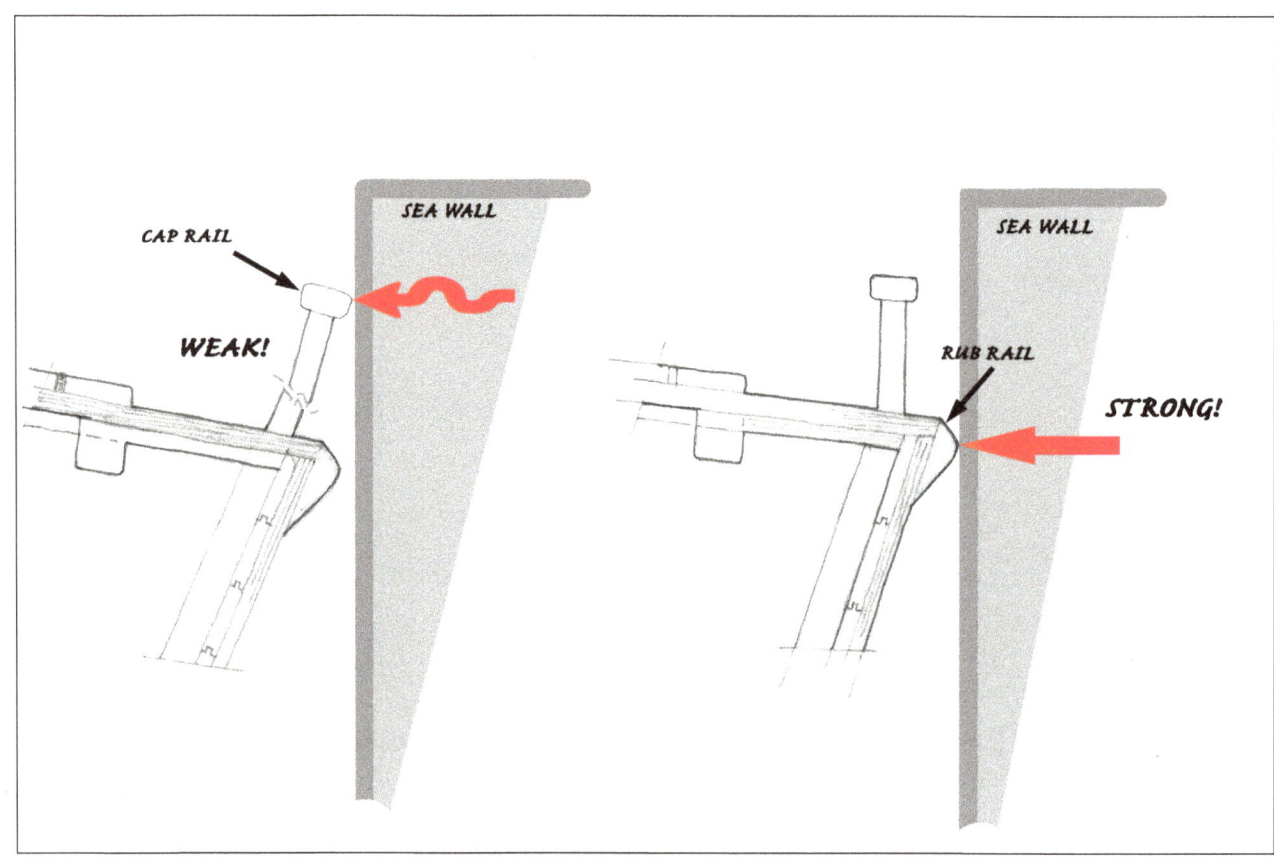

13-3 Vertical bulwarks. There is much to be said in favor of the rub rail, and not the cap rail, being the outer edge of the boat.

difficult and as bulwarks are vulnerable to collision, repairs will be all too commonplace. A better way is to set the bulwark inboard the thickness of the hull plating (see figure 13-1). Then the bulwark can be bolted through the deck without penetrating the hull plating. Any water that works through the bolt hole will evaporate into the cabin or drain away to the bilge.

Insetting the bulwark has an additional advantage. At least in the middle of the boat, where the bulwarks will be near-vertical, the inset will bring it inboard of the rub rail. When bumping against a pier or another boat, the rub rail, not the cap rail, will take the impact. The rub rail does its job of fending off minor collisions well, being placed at the corner of the hull/deck, an immensely strong point, and being built and finished both to absorb impact and to be easily replaced. The bulwark, with its fine cap rail, is cantilevered up from the hull/deck corner, a weak geometry, and is finished to grace the vessel, not to rub docks. It is vulnerable. A gentle collision will mar it and may well break it.

To make his bulwarks more robust, the builder, besides insetting them, may make them vertical all around the boat. He does this because even if the bulwark is inset an inch or two, as the topsides flare going forward, the bulwarks and its cap rail will flare outboard the rub rail. Again they will be the first thing to hit. The same is true of a stem casting overhanging the forward end of the deck, or of a taffrail overhanging the stern. Better to let the rub rail do the rubbing. Although it may take a bit of getting used to, I think vertical bulwarks make a nice look.

Waterways and scuppers

The surface formed by the inside of the bulwark and the outer edge of the deck is called the **waterway**. The waterway guides water running off the deck into the scuppers and overboard. Stopping the teak decking a few inches short of the bulwark gives the waterway a distinct gutter that channels the water. Dirt accumulates in this gutter, so it is a good idea to keep it free of hardware, allowing the water to flow freely and to make for easy cleaning. This requires care at the shrouds because the chainplates will intrude into the waterway.

If the waterways are painted white, they will show up at night as a distinct edge between boat and the deep, a nicety that will be appreciated by the crew. On the yachtiest yachts, scuppers drain the waterways via pipes that lead to thru-hulls below the waterline. This keeps runoff water from staining the topsides. Fine indeed, but the failure of one of these pipes may well sink the boat. The oceangoing sailboat is better safe than fine, so a better way is to provide an escape for the water directly overboard, under and through the bulwarks at their low points. Any scupper stains can be washed by an afternoon sail.

Chainplates

The chainplates transfer the very large tension load of the shrouds into the hull. In traditional construction the transfer is accomplished by bolting the chainplates to several frames, which then carry the load to the floors. With cold-molding, the load may be transmitted directly into the laminate, which disperses it through the hull. The Gougeon brothers developed the technique of imbedding studs in epoxy for fastening windmill blades to their hubs.[1] Their technique is an effective, inexpensive, and strong way to attach chainplates. Holes the diameter of the stud, or slightly larger, are bored in the desired location. The holes are partly filled with epoxy and the stud is inserted, embedding it with a surround of epoxy. Epoxy embedding prevents the sawing action described in figure 5-2. Used with stainless steel, to which epoxy adheres well, the epoxy prevents water intrusion. Epoxy -imbedded screws and studs are almost permanent. But

---

1 *The Gougeon Brothers on Boat Construction*, p. 301.

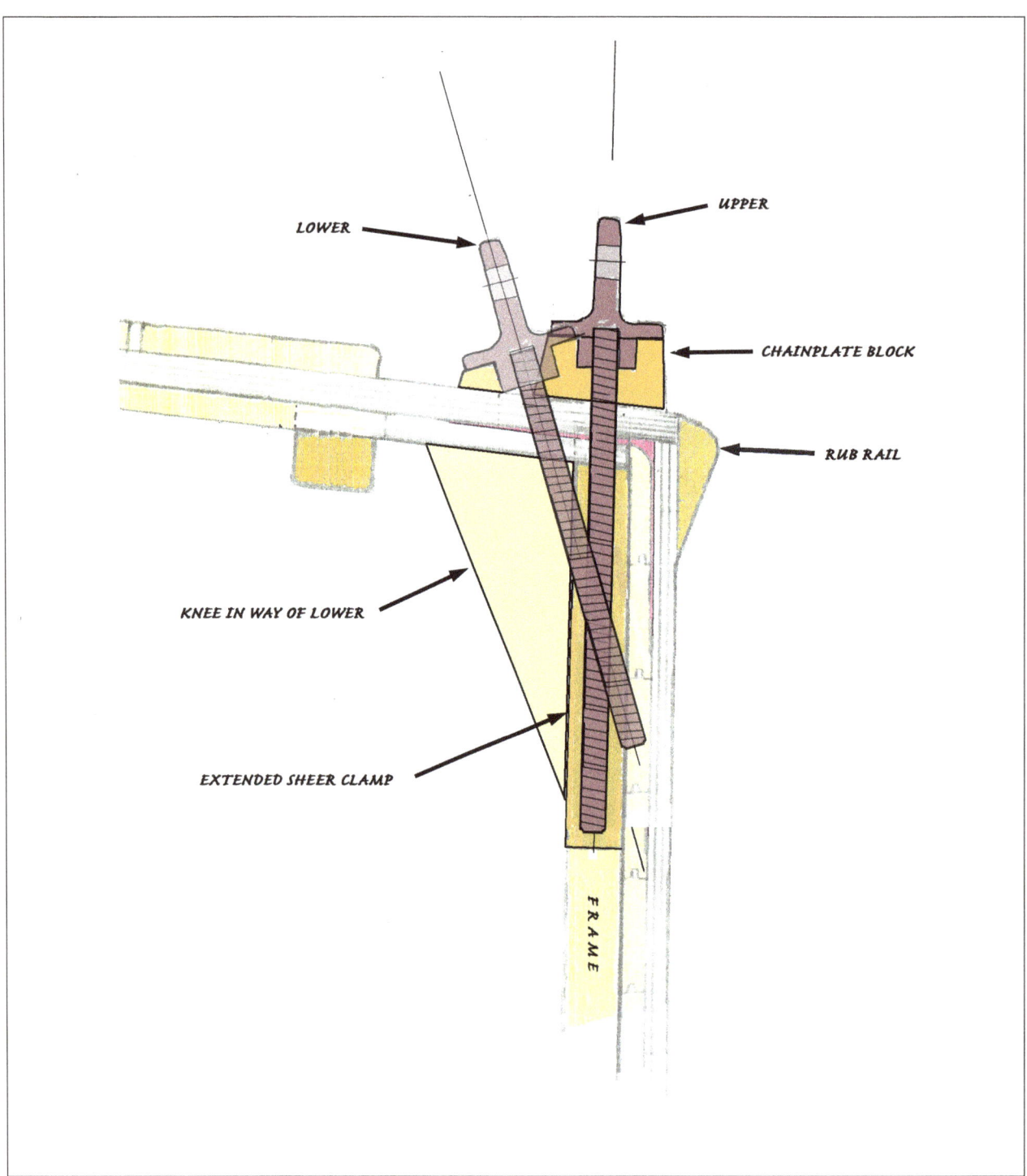

13-4 Chainplates. Chainplates have conflicting requirements. They want to be outboard, free of the bulwarks, and clear of the waterways. They are fastened with long studs embedded into the sheer clamp that need to be long, to not penetrate the laminate, and to inline with the shroud.

they can be withdrawn with heat. Heat liquifies the epoxy, transforming it from glue to lubricant, allowing the fastenings to be easily removed.

But such implants have a drawback if imbedded directly into the laminate. The hull laminate is thin, so the stud will be quite near the surface. There, it is susceptible to collision and exposure from the gradual sanding away of the hull as it is refinished year after year. Furthermore, the straining of the highly stressed studs may fracture the epoxy seal at the deck and allow water to enter into the plating. Any saltwater that works its way into the crack will cause anaerobic corrosion of the stainless studs and quickly break them.

Chainplates should be invulnerable to such petty attacks, so a better practice is to imbed the stud into the sheer clamp block. The sheer clamp in way of the chainplates should be big enough so that planking can carry the load on its cross grain in shear. As an example, on a forty-footer, the load on the upper shroud might be half of the displacement, or 10,000 pounds. Cedar planking has shear strength of about 200 pounds per square inch, so a gluing surface of about 50 square inches is required. A 2 × 6 inch block 12 inches long would suffice.

The shrouds are not always vertical. While upper shrouds pull nearly vertically, the lowers slope inwards; neither are perpendicular to the deck surface. To avoid side loads, the studs should be parallel to the shrouds they terminate. To correct the angle, a riser block is glued to the deck with its top correctly sloped. Underneath the deck, a vertical backing block may be glued to the sheer clamp to house and hold the stud. This block acts as a hanging knee. It reinforces the deck at a highly stressed point and is well worth its weight.

Since the studs run through the sheer clamp, directly under the bulwarks, the shroud terminations will necessarily interfere with the bulwarks. Advantage can be taken of this by using the riser block to interrupt and anchor the bulwarks and to raise the upper shroud terminals to the level of the cap rail.

Stanchion bases

Stanchions take high loads. When a boat is knocked down, water pressure will bend a 1 inch stainless steel tubular stanchion to horizontal. The same will happen when a man is hurled against stanchions. In addition, their bases are a good place for eyes to accept snatch blocks and snap shackles for temporary rigging attachments. Their loading is in tension, vertical and sideways. They should be kept clear of the waterways, where they will catch dirt and interfere with cleaning.

A good location for stanchions is along the edge of the teak decking at the inboard edge of the waterway. They can be bolted through the beam shelf stringer and a deck beam. If no attachment eyes are required, stanchions can be set directly over a stud epoxy-bonded directly into the beam shelf stringer. This will give a stanchion springing directly from the deck without any base at all. If the backing is sufficient, the stanchion will fail only by the destruction of the stainless tube, not by failure of the stud or the deck.

Rub rail

The rub rail is where a boat meets, at least above water, the hard things of this world. It wants to be strong, easily repairable, and finished to hide scuffs. It also wants to be on the outermost of the boat. As we have noted above, often it is the cap rail atop the bulwarks that finds itself—to its dismay—in this position.

I first became involved with the art of the rub rail when replicating Herreshoff's *ALERION*. On *ALERION*, as on many of his other boats, Herreshoff elegantly swelled the sheer strake to make a rubbing surface that stood proud of the surface of the hull. *ALERION*'s shear strake was made of oak about ¼ inch thicker than the lower planks. Herreshoff shaped its section into a smooth curve with no edges to catch. Bumping against a dock, its protrusion took the impact and protected the painted topsides below it. We developed Herreshoff's idea for the cold-molded hull of the *Alerion Class Sloop*. Not wishing to duplicate Herreshoff's exact shape, we exaggerated the swelling and pushed it to the top. The outer layer of four layers of planking on the *Alerion Class Sloop* is ¼ inch thick. Our sheer stock was molded from four-quarter material, so the swell projected ⁷⁄₁₆ inch from the general topsides, not enough to hang up on a dock edge but enough to distinguish us from Herreshoff.

While it is not an issue with a small day sailor like *ALERION*, for a larger cruising boat, using the sheer strake as a rub rail is not such a good idea. The rail should be sacrificial so that when it sustains damage, the damaged area can be easily cut away and replaced. The sheer strake itself is too important to be treated so casually. It is a difficult piece to remove for repair. The rub rail wants to be a sacrificial piece, independent of the general structure.

Another issue with using the sheer strake as a rub rail is that the strake is wider (the vertical dimension) than a rub rail needs to be. This requires it to be either cut to its spiled shape before molding its special section or else requires strong edge-setting to bend it to the curve of the sheer. Both make it a very difficult piece to repair or replace. The rub rail should be narrow enough to easily bend and twist to the sweep of the sheer and the flare of the bow.

An easy rail may be made of stock about twice as high as it is thick. Saw it diagonally across its rectangular section and round over the remaining right-angle corner. Then glue it to the upper edge of the hull (see figure 13-1.) A small boat can use 1 × 2 stock and a larger one 2 × 3. Such a rail is narrow enough that looks do not require it to be tapered. The shape will blend smoothly at its lower edge into the topsides, so it won't hang up on the dock.

Other topics

There are other items at the rail. A chapter could be spent on the design of the stem head discussing how the bow chocks might terminate the bulwarks and the rub rail. How does the headstay attach? the bow pulpit? Where do the running lights go, and how does their wire get back to the switch panel? How are the side chocks to be built? How will the taffrail be detailed?

There are other parts to a boat. Spars are an area rich in questions, some responding to epoxy bonding. There is a myriad of details concerning rig and rigging, electrical and pipe work, the selection of equipment. But I think we have gone into enough to get the big ideas across. It is time to bring this volume to a close with a summary of what we have accomplished, which I leave to the next brief chapter. And after that, an afterword for those who have old boats that they love too much to abandon without a fight.

CHAPTER 14

Conclusion

As usual, it took three tries to get it all right. *FANCY* was the first effort, *STARRY NIGHT* the second. The mental effort set out in chapters 8 through 13 is the third. Guided by the experiences recounted in chapters 2 and 3 and the science reported in chapters 4 through 7, we invented a process, described in chapters 7 through 13, for building an oceangoing sailboat suitable to the quest described in chapter 1.

Our third try answers questions raised by Sanford Boat Company's youthful gambit building the *Alerion Class Sloop*. The boat we have invented obeys the first five **Rules** of the sea and, under a proper master, may well obey the sixth.

We have used a number of innovations to accomplish our goal, and I list some of the prominent ones along with the names of those associated with them:

1. Epoxy bonding—Gougeon Brothers
2. Stringers to form and reinforce the hull shell— L. F. Herreshoff, Sanford Boat Company
3. Simplified stem—Sanford Boat Company
4. Light hull framing—Brad Pease of Pease Boat Works, Chatham, Massachusetts.

CHAPTER 14 CONCLUSION

5. Fiberglass taping around sharp corners— Sanford Boat Company
6. Plywood to replace thick, bolted timbers at structural corners: timber keel—Steve White of Brooklin Boat Yard; beam shelf— Sanford Boat Company
7. Accommodation mold—Sanford Boat Company
8. The bulkhead-block joint for installing interior accommodation—Sanford Boat Company
9. Longitudinal deck frame with light, cold-bent deck beams— Sanford Boat Company

No one of these novel techniques works alone; rather, all work in conjunction with each other to allow the construction of a boat that meets the **Rules**.

Another result of our quest was the discovery of economies in construction. My original discovery was the role of organization in the shop. As the *Alerion Class Sloop* project progressed, we worked out a method of building, documented it, sequenced it, and taught it to our crew. Doing so, we reduced our construction time by two-thirds.

But the *Alerion Class Sloop* remained an expensive boat, partly because of her exceptionally high specification and partly because her technology was new. Her project asked questions of economy; *FANCY*'s and *STARRY NIGHT*'s projects tried to answer them. Both *FANCY* and *STARRY NIGHT*, being the products of research projects, were expensive boats. But both their projects were fruitful. Accomplishing them, we discovered new, less costly ways of doing things. And while we were at work, the industry around us developed cost savers beyond our efforts.

Sanford Boat Company's major cost reductions:

1. Reducing the use of complexly shaped parts. Today, only the stem is hand-cut.
2. Using stringers to simplify shaping the hull.
3. Bending the framing over stringers and keeping them away from the accommodation to eliminate the need for twist or bevel.
4. Simplifying the keel/floor structure.
5. Simplifying the sheer clamp/beam shelf structure.
6. Framing the deck with lumber bent, rather than cut, to shape.
7. Installing the interiors with the accommodation mold technique.

And the big industry innovations:

1. Computer-aided lofting
2. Computer-controlled cutting and marking of plywood
3. Vacuum bagging
4. Vacuum infusion (still in its infancy)

While there is no yard using all these techniques, today, building an economical wooden sailboat for blue water sailors is feasible. I trust the time will come when it is commonplace. I offer this book to those builders and sailors who will bring it to reality.

I wish for you good luck and happy sailing.

AFTERWORD

I met an old acquaintance, a whaling-captain who said, "Come to Fairhaven and I'll give you a ship. But," he added, "she wants some repairs."

. . . The people of Fairhaven, I need hardly say, are thrifty and observant. For seven years they had asked, "I wonder what Captain Eben Pierce is going to do with the old *SPRAY*?" The day I appeared there was a buzz at the gossip exchange: at last someone had come and was actually at work on the old *SPRAY*. "Breaking her up, I s'pose?"

"No; going to rebuild her." Great was the amazement. "Will it pay?" was the question which for a year or more I answered by declaring that I would make it pay.

Joshua Slocum, *Sailing Alone Around the World*

CHAPTER 15

Restoring Traditionally Built Craft

EVEN IF COLD-MOLDING weren't the preferred technique for building new boats, it would be important as a technique for restoring traditionally built wood boats that have deteriorated from use and age. Old boats are valuable. Some are very valuable. Their value stems not only from their beauty and history but also, more fundamentally, from their design. Based on knowledge of the sea generated over centuries of experience, they sail better, are more sea-kindly, and are more seaworthy than many of their modern sisters. That knowledge, forgotten or eschewed today, is built into their bones and makes them worth saving.

But how to restore them? Contemporary practice is, dismantle the relic and build it new by replacing each old piece with new material newly fastened. Assuming the repairers avoid the temptation to expand her machinery with all the "modern inconveniences," the result could be an historic interior (if it is not replaced also) surrounded by a sterile new hull. Lost will be all her marks, her history, and her soul.

Another way of restoration is to use cold-molding to restore strength to her tired hull. Laminating two or three layers of thin veneer over the old refastened planks weakly held together with broken frames, covering her leaky seams, and renewing her hoop strength will bring her back. Then the interior, of both her hull and her accommodation, will remain as is, with the beauty marks and patina of age intact, to continue sailing to new adventures. The cold-molding will stop her leaking, shaking, and rattling. It has an indefinite life, costs about 10% of a new hull, and, best of all, lets you keep your boat. You don't have to abandon an old friend.

Cold-molding by itself cannot fix everything. It will not restore heavily rotted planks or rotten centerline structure. But it will cure loss of hoop strength, broken frames, loose fastenings, open garboards, and other failings of metal fastenings. These are the common failings of traditionally built wood boats, and many lovely boats on their way to becoming wrecks may be restored with cold-molding.

CURLEW

Cold-molding made possible Tim and Pauline Carr's spectacular voyages around South Georgia with *CURLEW*. They found *CURLEW* in Malta in 1967. She was a 70-year-old, abandoned Falmouth Quay Punt in almost useless condition. The Carrs cleaned her, got rid of her engine, and sailed her for 16 years, around the Mediterranean and on to New Zealand, repairing her as they went. In the process they grew to love her. They learned from her how much contemporary sailors had forgotten.

But by age 90, *CURLEW* had problems conventional repairs could not easily correct. *CURLEW* had been exceptionally well built, but, no matter, her fastenings were rusting away and damaging her frames as they did. It is, indeed, a tribute to the Carrs' seamanship that they kept her seagoing for as long as they did. With her planks no longer secured to her framing, *CURLEW* was no longer structurally a shell; rather, she was a loosely attached bundle of sticks. And weakened sticks (her planks had suffered more than one refastening) at that. Without shell integrity she could not stand up to her working loads.

To the Carrs, *CURLEW* was too dear to discard, so they decided to restore her strength with cold-molding, a technique about which they had recently learned. By epoxy bonding a thin laminate over her weakened planking, the Carrs got her planks working together again as a structural shell. She regained her old strength. They saved her decks, her interior, her gear, her quirks and marks—most important, her soul. All they had lost was the fear that she might break up and sink.

After showing her prowess by beating the new designs in proper regattas, the Carrs sailed *CURLEW* through the wind and waves of the

15-1 ***CURLEW* in South Georgia.** The Carrs fixed up *CURLEW* over a period of 16 years as they learned to know and love her. Their restoration culminated in cold-molding her hull to bring back the strength lost by fastenings turned to rust. After cold-molding she was ready to explore the world's stormiest ocean with her beauty, performance, and soul intact.

Roaring Forties and Furious Fifties to South Georgia. There they explored the islands and gave the world their beautiful pilot, chronicle, and natural history, *Antarctic Oasis*. It should be in your library.

The Carrs also gave *CURLEW*, well into her second century, to England's National Maritime Museum in Falmouth, where she is exhibited afloat. It is a shame that the Carrs' could not have found another sailor to keep her voyaging, but times have changed. Such sailors are hard to find. So *CURLEW* is well placed at the museum, where she is a stubborn relic attesting to the great punts created by Falmouth's watermen and to the pluck and skill of certain English sailors who can take a small inshore work boat and use her to explore the wildest oceans of the world.

IMPALA

Entirely independently of *CURLEW* and the Carrs, in 1986, I came to own *IMPALA*. She is a 56 foot yawl that I had first met in 1955, at age 12, as she was being ogled from shore by my elders. They considered her the most beautiful of all Olin Stephen's creations. By the time I came to own her, she was 34 years old and still extraordinary. But even though she was very well built by Abeking and Rasmussen, she had taken some hard knocks and suffered some incompetent care. She was in less than perfect condition.

Like the Carrs, I sailed *IMPALA* a bit before I started repairing her and, like them, I first used traditional repairs for her traditional structure. I replaced broken pieces with new material using her original construction methods. In 1987, Brooklin Boat Yard of Maine replaced a good portion of her bottom planking where she had been damaged by electrolysis and also by a mysterious grounding reputedly suffered while drug-running in the West Indies. Brooklin's repair bought me some time.

As a traditionally built boat ages, she begins to leak. Deck leaks are the worst. Hull leaks will only sink the boat, and before they do, their waters rest in the bilge, where they may be pumped overboard. Deck leaks create slow misery by dripping on the crew as they make their way to the bilge. *IMPALA* had a very leaky deck.

IMPALA's decks had worn thin. Originally 1¼ inches thick, walking and scrubbing had reduced their thickness by about ⅜ inch. The fastening bungs were coming loose, which would soon allow water into the heart of the deck beams. That would cause them to rot away, destroying the entire structure. Rather than remove the deck planks and replace them with new ones of the original thickness, in 1989, Brooklin Boat Yard renewed her decks a different way. Over the existing deck they laid down a new layer of teak decking, glued to the old. This covered the leaking bungs and sealed the fastenings safely away from intruding water.

15-2 *IMPALA* leaving Nantucket for Gibraltar, May 2011. My experience with *IMPALA* was similar to the Carrs', although *IMPALA* was bronze fastened and only 44 years old compared with *CURLEW*'s 85. *IMPALA*'s hull was tired when I fixed her in 1998. But not after cold-molding. Here she is leaving Nantucket for Gibraltar in 2011 on her second voyage to Europe since being remade strong. We left in a moderate gale on a Thursday to avoid leaving on a Friday the 13th.

As is so often the case, repairing an old wood boat is like an archeological dig. The deeper you go, the more you find. While re-decking *IMPALA*, we discovered previously unseen problems. Over the years, her lower hull had failed. Her mast had pushed her maststep down about ¾ inch, taking her deck with it. This caused a dimple in her deck. We repaired the dimple as we resurfaced the decks, but the deck repair did nothing for the frames that the mast had broken on its way down. As discussed about *PIERA*, in chapter 2, this is a classic failure of traditionally built boats carrying tall Marconi rigs—insufficient hoop strength to carry the large tension loads imposed by mast and shrouds.

Repairing this damage by traditional means seemed fruitless because the damage resulted from the inadequacy of traditional construction. Replacing the broken pieces would not make the inadequate adequate. I was intrigued by the idea of using cold-molding to eliminate the original weakness. So, during the winter of 1990–1991, Generation III Boatyard of Cambridge, Maryland, laminated two layers of ⅛ inch mahogany veneer over *IMPALA*'s bottom, in way of the mast step. The laminate ran from a little below the waterline to the keelson for a length of about 10 feet. The laminate was sanded to smoothly taper out to zero thickness at its edges.

I watched the repair carefully over the next six years. I had expected it to crack where it tapered out to nothing, but, to my surprise, it did not. Examining the hull out of water, all one could see was a smooth surface under the mast with the plank seams suddenly appearing beyond the extent of the repair. The mast step was stabilized, and *IMPALA* stopped leaking. There was no need to change the mast step; it was a bronze weldment and, although at the scene of the failure, it, itself, was not part of the failure. The problem was the frames, which were, and always had been, insufficient to take their loading. The laminate took over their job and made them redundant.

But *IMPALA* was getting older; by 1998, she was 44. Although her bronze screw fastenings were in good shape, her frames were not. Many had been broken—in the tight reverse curve of her garboard area and across her bilge stringer (see figure 4-11 B). Her hull was, as they say, getting soft. When a wave hit her, the whole boat shimmied and shook, making a noise like the slamming of an old jalopy door. All the while, I was using her for deep sea work. She was progressively weakening with each wave that hit her and each year that passed. I wanted her stronger.

Of course I could have rebuilt her the old way, but I did not want to do that, not only because of the expense, but because I knew that if I did, in another thirty years, I would be right back where I had started with a weak boat and a tired hull. More than that, I knew that after the rebuild, most of my old boat would no longer be with me—she would be on the refuse pile. I would have thrown away wonderful

IMPALA with all her history and adventures. She would no longer have her beauty marks of half a century of sailing. No old frames and deck beams, just shiny, new, perfectly fitted, unremarkable wood. No! I loved my boat; I did not want a new one. What I wanted was the old one made strong.

So motivated, and encouraged by the success of the earlier repairs, I engaged Pease Boat Works of Chatham, Massachusetts, to laminate two additional layers of ⅛ inch veneer over the entire bottom. Unsure of what diagonal laminated topsides would look like, the laminations stopped at the top of the boot stripe. Where the frames were broken above the waterline (generally broken by the bilge stringer), in the forepeak and in the counter, Pease laminated "sisters" directly to the inside of the planking, ½ inch thick and 8 inches wide, clear of the frames by about an inch on either side. IMPALA went back in the water with a new paint job and sailed the 1998 Bermuda Race a few weeks later. She had a new sound. When a wave smacked her hard, it did so with a muffled thump that reminded one of the door slammed on a new Mercedes.

It is remarkable how little material it takes to mend a broken hull. IMPALA weighs 52,000 pounds. The laminate applied to her hull weighed about 450 pounds. It also added about 800 pounds of buoyancy, so she floated a little higher. Her thicker hull increased her waterline about 2 inches. Repaired she was a slightly bigger boat, slightly less burdensome, and slightly faster. Best, she was rock solid and no longer leaked.

Twenty-one years later, in 2019, after three Atlantic crossings, IMPALA's hull shows no signs of deterioration. Her repair has been long lasting—and reassuring. Further, when visitors come below, I am proud to show them Abeking's original work, not some 21st century imitation of it. In my bunk at sea I look up at the underside of her 65-year-old varnished teak deck, more beautiful each year and still going strong.

Cold-molding is a way to "save" an old boat without destroying it.

APPENDIX A

Reading the Tape

TAPE MEASURES in America will read in inches. The inches are divided into sixteenths via a series of marks for halves, quarters, eighths, and sixteenths. It is a great system except for two things. When reading the tape, one's eye tends to round off to the nearest eighth, quarter, half, or even whole inch. This not only introduces inaccuracies, it also instills doubt. One might round to an eighth one time and maybe a half the next; one will unlikely remember which he did when. Also, math is hard; it is difficult to add and subtract fractions of different denominators.

These difficulties were introduced by our fifth-grade arithmetic teachers, who taught us that fractions should be reduced to the smallest possible denominator. Being good students, we have done as she told us and ended up with five different classes of measurements from one tape measure. We can read in whole inches, halves, quarters, eighths, or sixteenths. Our teacher also taught us that these different fractions could be added and subtracted by just changing them all to the lowest common denominator,

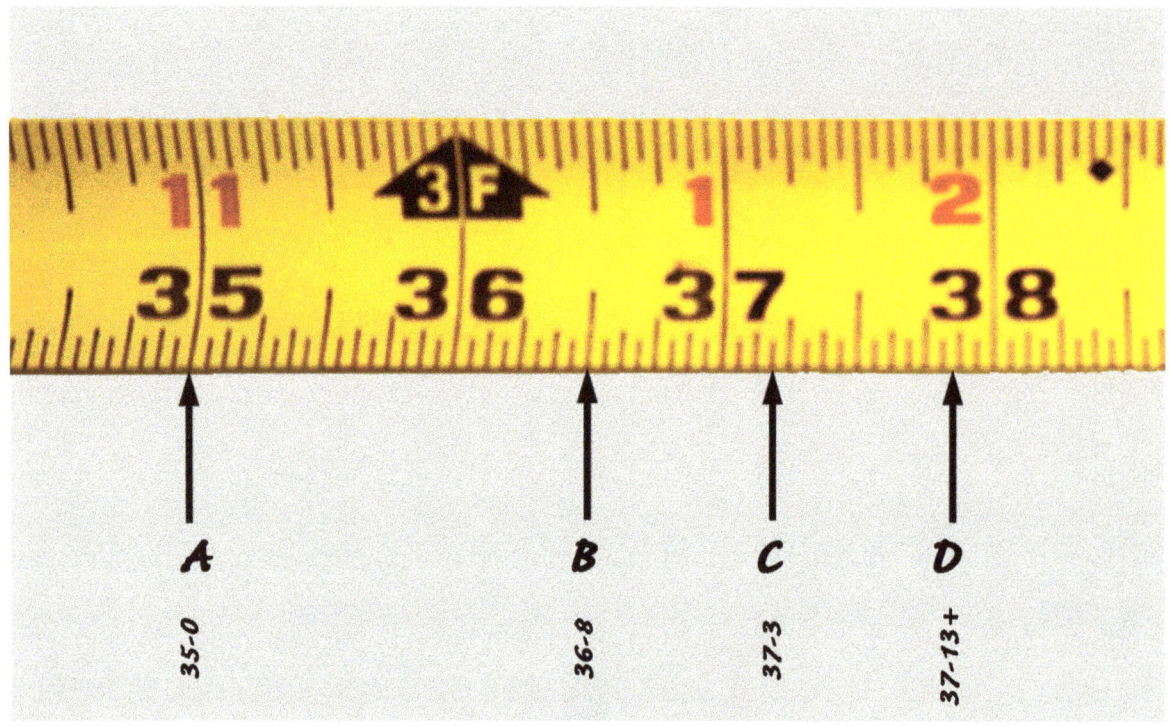

A-1 How to read the tape.

adding the numerators. Then, having arrived at an answer, we could put them back into presentation form by reducing the numerator and denominator of common factors to achieve the minimum denominator. Remember?

There is a way around these difficulties. Respect our teachers, but do all the readings in sixteenths. Read ½ as ⁸⁄₁₆, ¾ as ¹²⁄₁₆, ⅝ as ¹⁰⁄₁₆. Since we have decided to do everything in sixteenths, we know that 10 means ¹⁰⁄₁₆ so we can drop the sixteenths altogether. Using this sixteenths system, 15¾ inches becomes 15 ¹²⁄₁₆, which shortens to a spoken "fifteen-twelve" and is written "15-12." Notice, "fifteen-twelve" is easier to say than "fifteen and three quarters," three syllables rather than six, and in the sixteenths system the verbiage does not have the aliterative difficulties that "eighths" and "sixteenths" cause.

Our new numbers are not only easier to enunciate, they are easier to add and subtract. Using sixteenths for all numbers eliminates two major steps in the arithmetic: there is no need to find a common denominator (it is always 16), nor is there need to reduce the denominator of the result—it remains 16. I'll just ask you and you can answer for yourself: which is easier, to add ¼ inch to 1 ⁵⁄₁₆ inches, or to add 0-4 to 1-5 (1-9)? With subtraction, would you prefer to subtract $5/16$ from $1\,1/2$ inches, or to subtract 0-5 from 1-8 (1-3)?

As for the temptation to round off, when reading the tape using the sixteenths system, you will ask yourself, "Fifteen inches and how many sixteenths?" The answer is a whole number from 0 to 15. There will be no temptation to round $9/16$ to $8/16$, because they are both equally easy. You will take pride in the precision, and more, the avoidance of approximation will bring you ease—doubt and uncertainty go away.

Most people, after they try using the sixteenths system, like it and never go back to eighths, quarters, and halves. But in some delicate cabinetry work, a sixteenth is not a fine enough measurement. If you try to switch to the next level of precision and use thirty-seconds, the numbers get too big to be perceived, or remembered. Stay with sixteenths. If you need to measure to a thirty-secondth, add a plus sign to the smaller sixteenth. In other words, for $1^{27}/_{32}$ write 1–13+. Read it "one-thirteen plus." It is easy to write, easy to add, and subtract, and, most importantly, easy to visualize. Try the sixteenths system. I think you will like it.

APPENDIX B

Drawings and Dimensions

SAILBOATS ARE complex objects and their construction involves two tasks, related yet fundamentally different. First the builder fashions parts, and second he assembles them into a boat. The two physical tasks involve two intellectual tasks that actually develop in the opposite order. First the designer conceives of the vessel as a functioning whole; then he designs the elementary parts that make it up. These two tasks result in two sets of drawings, the **design drawings** that show how the parts come together to make the functioning whole, and the **shop drawings** that give the builder instructions for making the elementary parts. As an example, the *Alerion Class Sloop*, a very well-documented design, uses about 60 large-format 22 inch × 34 inch design drawings and about 160 smaller 11 inch × 17 inch shop drawings, each showing how to make one or more parts.

Not surprising, shop drawings and design drawings use different dimensioning techniques. Dimensioning is used to add numerical information to the graphic. The advantage of using numbers is that, unlike the pencil line, they can be made as precise

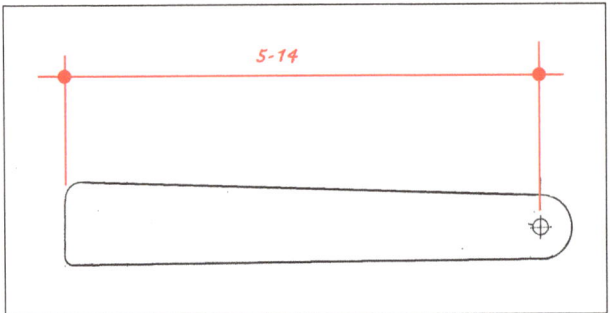

B-1 Relative dimension. Three red lines, two dots, and a number give the precise distance of the hole center from the left end of the door clapper. It will take an additional set to position it vertically.

as desired, and they are easily manipulated by arithmetic. They make for easy communication between drawing and the craftsman's tools, his measuring tape, table saw, and drill bits.

Shop drawings mostly use a conventional dimensioning system that indicates the distance between two selected points. A composite symbol composed of three lines, two shapes (arrow heads or balls), and a number identifies the points in question and gives the distance. I call such dimensions **relative dimensions** because they tell you where one point is relative to another. Relative dimensioning works well for the layout of a simple part. With simple parts, the draftsman can define each key point, saving the craftsman the need to do any calculation.

But for design drawings, relative dimensions have disadvantages. They offer no positional information. They define what is to be made, but they do not show where it is to go, nor how it fits together with the other parts to make the complex whole, which is the job of the design drawings. Design drawings contain many more relationships needing definition than can be shown by relative dimensions without cluttering the drawing beyond readability. This is what happens with an airline flight map where all the flight lines coalesce into a confusion of lines at major hubs. No particular route is traced and all the diagram manages to show is that the system is complex. Good PR but not very useful.

Another disadvantage of relative dimensions is that, being relative, if one is in error, the error may begin a cascade of errors that mis-dimension other points associated with it. For instance, if the dimension highlighted in red in figure B-2 were incorrect, the notch would be mis-positioned, and the surface indicated with

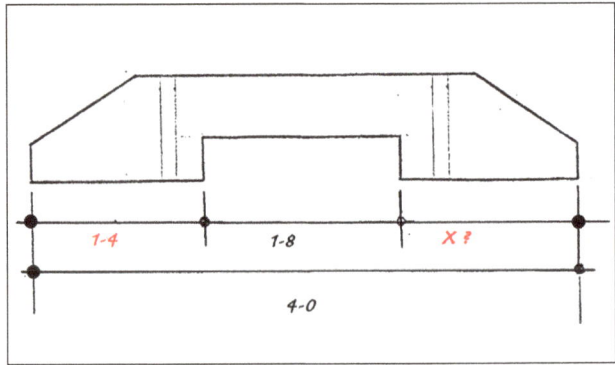

B-2 Relative errors. Chaining relative dimensions saves space and tick lines but runs the risk that an error in one dimension will generate errors in others. On this sketch, if the 1–4 inch dimension in red is in error, the notch is mislocated and dimension X is incorrect as well. With a long chain, the errors cascade.

APPENDIX B — DRAWINGS AND DIMENSIONS

the red arrow would be the wrong size. The longer the chain of dimensions, the more errors.

Design drawings need another dimensioning system. It was invented by Rene Descartes, way back in 1637, as a way to numerically fix a position on a graphic. In an effort to bring measurement and algebra to Euclid's geometry, he came up with the idea of a **coordinate system**, now called Cartesian, to fix the position of any point of interest. Descartes established on the graphic plane a reference system consisting of two perpendicular axes. The point where the axes crossed he called the origin. Then he defined the coordinates of a point as its distance from each of the two axes. The position of any point is fixed by the two numbers of its coordinates. Applying Pythagoras' theorem to the coordinates of any two points, he could calculate the distance between them. Once the coordinates were established, all relative dimensions could be calculated at will. This is how a computer organizes points in a design and why it can produce relative coordinates instantly.

Descartes' system may be extended to three dimensions to locate points in space, and that is what we do for a boat design. Sailboats are

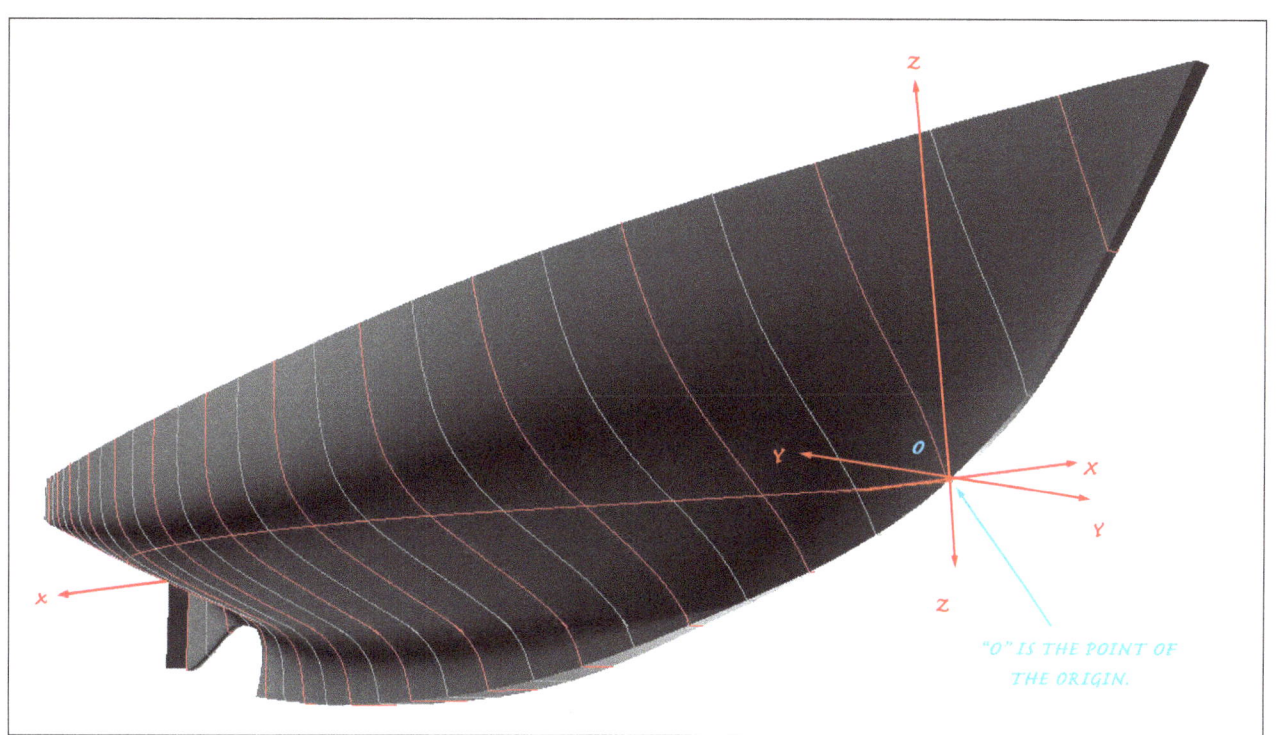

B-3 **Axes and Origin.** Cartesian dimensions use a set of axes and an origin to locate any point with a set of coordinates. Here the axes X, Y, and Z, for three dimensions of STARRY NIGHT, are pictured on a surface model of her hull. The point O, where the three intersect, is the origin.

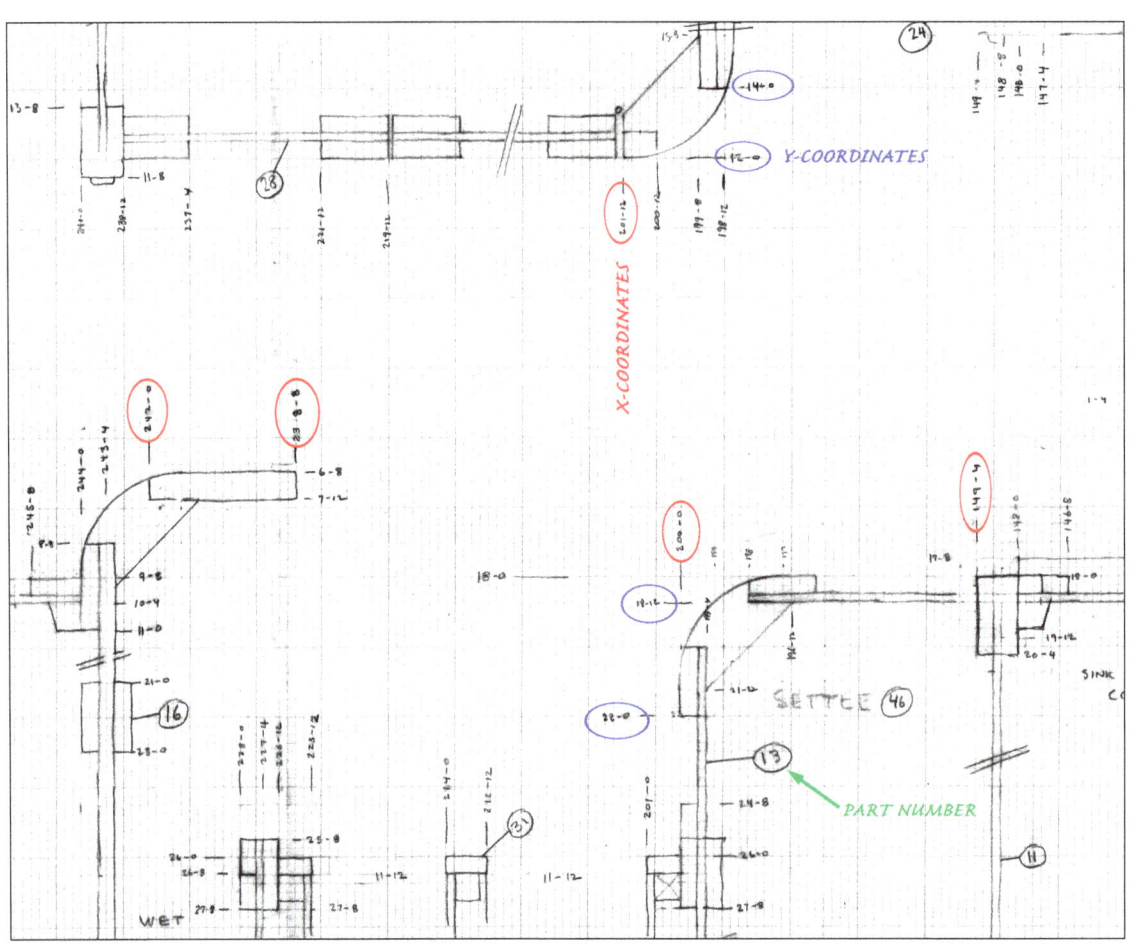

B-4 Cartesian coordinates. This is a blowup of a section of a larger drawing that lays out *STARRY NIGHT*'s entire interior accommodation, looking down from above. Because the structure is parallel to the axes, each point is determined by a single number; the vertical numbers are the X coordinates and the horizontal numbers are the Y coordinates. Using subtraction, the widths and lengths of the pieces and their parts can be calculated precisely.

three-dimensional. We need three axes. From them we get three coordinates. The choice of coordinate system is arbitrary; the technique works no matter which coordinates you choose. But some choices make calculations much easier than others. On my designs I use the centerline of the waterplane as the X axis, which corresponds to length. Perpendicular to it is the Y axis also in the waterplane, and it corresponds to width. Finally, the Z axis is perpendicular to the waterplane and corresponds to height.

I often chose the origin, or zero point, at the forward waterline ending. This choice makes the X coordinates of points near the bow negative, which some may not prefer. Herreshoff's *ALERION* had frames spaced 9 inches apart. For the *Alerion Class Sloop* we chose as origin a point 9 inches forward of frame #1; this would be the position of frame #0 if there were such a frame. The point was actually a couple of inches out in space forward of her bow. Odd, but using that origin point, all of her X coordinates were positive. As Descartes would say, it does not really matter; the origin is chosen for convenience.

With the origin at the waterplane, points above the waterplane are positive, and the ones below are negative. The Y axis is a little trickier; it makes sense to have the origin on the centerline, but which side is to be positive and which negative? I, personally, drop the sign. If there is ambiguity, I use "port" and "starboard." Hulls are usually symmetrical around the center line, so conventionally only one side is drawn and there is no issue. But accommodations and equipment are often not symmetrical, so sometimes you must make a distinction.

Once the Cartesian coordinate system is set up, each point of interest can then be numerically defined (dimensioned) by using its coordinates. The system is analogous to latitude and longitude for locating points on the earth's surface. The lat/long not only gives you the location of a port, the distance to any another port may be determined from that other port's lat/long using the appropriate formulas.

BIBLIOGRAPHY

VOYAGING

A selection of great ocean voyages in modest yachts, including a few volumes on the appropriate vessels for doing so.

Belloc, Hilaire, *On Sailing the Sea*, selections made by W. N. Roughead, Rupert Hart-Davis, London, 1951.

Belloc was a historian, theologian, and political philosopher who wrote about life and sailing, especially before World War I.

Bradford, Ernle, *Ulysses Found*, 1963, Harcourt Brace & World, Inc., New York, 1964.

In an act of detective work, Bradford retraces via his sailboat Ulysses' track around the Mediterranean from Troy to Ithaca. Before Bradford, many thought Ulysses a fictional character.

Carr, Tim and Pauline, *Antarctic Oasis: Under the Spell of South Georgia*, W. W. Norton & Company, New York, 1988.

The Carrs capped a life of cruising with a voyage to South Georgia. They took charge of the whaling museum there and over a period of years created this extraordinarily beautiful pilot and natural history of those extraordinary islands. Their 122 year old Falmouth Quay Punt, *CURLEW*, is discussed in chapter 15 of this volume.

Chichester, Francis, *The Lonely Sea and Sky*, Coward McCann, Inc., New York, 1944.

Chichester became famous by winning the first single-handed transatlantic race. He was also an aviation pioneer and a fine writer. This volume contains some of his early writing about air and sea adventures.

Childers, Erskine, *The Riddle of the Sands*, 1903, Mariner's Library, Rupert Hart-Davis, London, 1955.

The finest sailing mystery ever written and one of the best descriptions of navigating small craft in shoal waters. A pleasure.

Dumas, Vito, *Alone through the Roaring Forties*, Adlard Coles Limited, London, 1960.

A unique single-handed circumnavigation in a 31-footer through the world's wildest oceans. Done entirely below 35°S while the northern world was at war, it is a shame to paraphrase the story told so well by Dumas himself. Obtain a copy and read it!

Howard, Henry, *The Yacht Alice: Planning and Building*, Charles E. Lauriat Company, Boston, 1926.

———. *The Yacht Alice Twenty Years After*, Dodd, Mead & Company, New York, 1946.

Howard describes his *ALICE*, an unusual design, in great detail and includes an account of sailing her on a long voyage. In his second volume he discusses her again after twenty years of experience with her. The discussion of a vessel used over a long period is rare and useful.

Johnson, Irving, *The PEKING Battles Cape Horn*, Milton Bradley Company, Springfield, MA, 1932.

———, with Electa, *Westward Bound in the Schooner YANKEE*, W. W. Norton, New York, 1936.

———. *Sailing to See*, W. W. Norton, New York, 1939.

———. *YANKEE's Wander World*, W. W. Norton, New York, 1949.

———. *YANKEE's Peoples and Places*, W. W. Norton, New York, 1955.

———. *YANKEE Sails across Europe*, W. W. Norton, New York, 1962.

Irving Johnson learned deep sea sailing in 1929 at the close of the age of sail by making a three-month voyage from Hamburg to Chile aboard the four-masted steel bark *PEKING*. Besides reading his book, one must see the movie made by Johnson with a 16mm camera.

A few years later he and his bride Electa purchased their first *YANKEE* and created the concept of exploring the world aboard a yacht using a paying college-age crew to run the ship. Johnson's tradition lives today with the *PICTON CASTLE*, captained by Dan Moreland, who learned his trade aboard *ROMANCE*, captained by Art Kimberly, himself a Johnson protégé from *YANKEE*. The writings are good, the pictures better, and the seamanship best of all.

BIBLIOGRAPHY

Mitchell, Carleton, *Passage East*, W. W. Norton, New York, 1953. Racing *CARIBBEE* from Bermuda to Plymouth, June 1952.

A beautiful book by America's greatest ocean racer, who claimed he had cruised ten miles for every one raced.

Moitessier, Bernard, *The First Voyage of the Joshua*, 1969, William Morrow & Company, New York, 1973.

———. *The Long Way*, Doubleday and Company, New York, 1975.

The single-handed, around-the-world race of 1968–1969.

———. *Cape Horn, the Logical Route*, Granada, London, 1969.

———. *A Sea Vagabond's World*, Sheridan House, Dobbs Ferry, NY, 1988.

———. *Sailing to the Reefs*, Sheridan House, Dobbs Ferry, NY, 2001.

Moitessier, a French colonial from Vietnam, expresses sailing as a spiritual approach to life. He is one of the greats.

O'Brien, Conor, *Across Three Oceans*, Rupert Hart-Davis, London, 1949.

———. *From Three Yachts*, Rupert Hart-Davis, London, 1950.

Description of a voyage around the world in *SAOIRSE* in 1923–1925.

O'Brien was an Irish yachtsman and gun runner. He was an early and important commentator on what was required of the proper ocean sailing yacht.

Pidgeon, Harry Clifford, *Around the World Single-Handed: The Cruise of the "Islander,"* Mariners Library #12, Rupert Hart-Davis, London, 1951. First published 1933.

———"Around the World in the *Islander*," National Geographic Magazine, February 1927 pp 141-205

Harry Pidgeon's account of the second single-handed cruise around the world, in his homemade boat, was very popular at the time. It is a good story by an able man.

Robinson, William A., *10,000 Leagues Over the Sea*, Harcourt Brace and Company, New York, 1944. First published 1932.

Robinson's account of sailing 32-foot, Alden-designed *SVAAP* around the world, mostly with a paid hand aboard.

———. *Voyage to Galapagos*, Harcourt Brace and Company, New York, 1936.

Story of Robinson sailing *SVAAP* to the Galapagos Islands, developing appendicitis, a miraculous rescue, and loss of *SVAAP*.

———. *To the Great Southern Sea*, Harcourt Brace and Company, New York, 1956.

———. *Return to the Sea*, John de Graff, Inc., Tuckhoe, NY, 1972.

Robinson's story of sailing his great *VARUA* from Tahiti to southern Chile up to Panama and back to Tahiti to test her against the southern ocean. *VARUA* was specially designed for such a voyage by Starling Burgess and Francis Herreshoff. *VARUA* is interesting for this book as she was composite construction, wood shell over steel frame, an effort to introduce sufficient "hoop" strength into traditional construction.

Slocum, Joshua, *Sailing Alone around the World*, The Century Company, New York, 1900; Penguin, with introduction by Thomas Philbrick, New York, 1999.

The brilliant book by the man who started it all.

Smeeton, Miles, *Once Is Enough*, Rupert Hart-Davis, London, 1960.

An amazing story of the Smeetons headed from the South Seas to Engand via Cape Horn. *TZU HANG* was well built and strong, but caught in a great gale, their 45 foot ketch pitchpoled, broke off her deck house, and came close to sinking. They limped into Chile, repaired the damage, and continued, only to have a similar experience again. John Guzzwell of *TREKKA* fame was aboard for part of the trip and used all of his carpentry skills to help keep the ship afloat.

Snaith, William, *Across the Western Ocean*, Harcourt, Brace & World, Inc., New York, 1966.

Two transatlantic passages, one the 1963 ocean race Newport to Eddystone Light aboard his yawl *FIGARO*.

Tilman, H. W., *In MISCHIEF's Wake*, Hollis & Carter, London, 1971.

Tilman used yachts converted from Bristol Channel pilot cutters to voyage to Arctic and Antarctic islands for the purpose of climbing their mountains.

DESIGNS AND DESIGNERS

In recent years the work of almost all the important designers of the 20th century has been documented by either biography or autobiography. Here is a selection of works, with the designer's name preceding each work.

John Alden
Carrick, Robert, and Henderson, Richard, *John G. Alden and His Yacht Designs*, International Marine Publishing Company, Camden, ME, 1983.

Colin Archer
Leather, John, *Colin Archer and the Seaworthy Double-Ender*, International Marine Publishing Company, Camden, ME, 1979.

William Atkin
Atkin, William, *Of Yachts and Men*, Sheridan House, New York, 1949.

Dick Carter
Carter, Dick, *Dick Carter, Yacht Designer*, Seapoint Books + Media LLC, Brooklin, ME, 2018.

S. S. Crocker
Crocker, S. Sturgis, *Sam Crocker's Boats*, International Marine Publishing Company, Camden, ME, 1985.

Laurent Giles
Lee, Adrian, and Philpott, Ruby, *Laurent Giles and His Yacht Designs*, International Marine Publishing Company, Camden, ME, 1991.

Maurice Griffiths
Griffiths, Maurice, *Dream Ships*, Conway Maritime Press and International Marine Publishing, Co., London, Camden, ME, 1974. First published 1949.

———. *Maurice Griffiths, Sixty years a Yacht Designer*, Conway Maritime Press Ltd., London, 1988.

BIBLIOGRAPHY

Francis Herreshoff
Taylor, Roger C., *L. Francis Herreshoff, Yacht Designer, Vols. 1 and 2*, Mystic Seaport, Mystic CT, 2015, 2019.

Nathaneal Herreshoff
Herreshoff, L. Francis, *Capt. Nat Herreshoff, The Wizard of Bristol*, Sheridan House, New York, 1953.

Ted Hood
Hood, Ted, *Through Hand and Eye*, Mystic Seaport, Mystic, CT, 2006.

Ray Hunt
Grayson, Stan, *A Genius at His Trade*, Old Dartmouth Historical Society, New Bedford Whaling Museum, New Bedford, MA, 2015.

Ralph Munroe
Gilpin, Vince, *The Commodore's Story*, Livingston Publishing Company, Narbeth, PA, 1966. Originally published 1930.

Aage Nielsen
Bray, Maynard, and Jackson, Tom, *Worthy of the Sea*, Tilbury House, Gardiner, ME, 2006.

Philip Rhodes
Henderson, Richard, *Philip L. Rhodes and His Yacht Designs*, International Marine Publishing Company, Camden, ME, 1981.

Olin Stephens
Stephens, Olin, *All This and Sailing Too*, Mystic Seaport, Mystic, CT, 1999.

———. *Lines*, David R.ß Godine, Boston, 2002.

Albert Strange
Clay, Jamie, and Miller, Mark, *Albert Strange on Yacht Design, Construction and Cruising*, The Albert Association, Woodbridge, UK, 1999.

YACHT DESIGN

A selection of books on yacht design by scientists, artists, and sailors.

Barnaby, Kenneth C., *Basic Naval Architecture*, Hutchinson, London, 1949.

A nice volume that will give a sailor enough naval architecture to understand how and why his boat moves through the water.

Bavier, Robert N., Jr., *Faster Sailing*, Dodd, Mead & Company, New York, 1954.

A good introduction to the speed potential of light displacement and rating rules by the inventor of the One-of-a-Kind Races.

Butler, T. Harrison, *Cruising Yachts: Design and Performance*, Robert Ross & Co. Ltd., London, 1945.

A short treatise on yacht design by an English doctor who produced a series of small, tidy cruisers.

Chapelle, Howard, *Yacht Designing and Planning*, W. W. Norton, New York, 1936.

———, [Chapelle], *American Small Sailing Craft*, W. W. Norton & Company, Inc., New York, 1951.

Howard Chapelle spent the Depression studying and documenting the sailing ships that were then becoming extinct. He presented his findings in a series of books. In the process he wrote this "how to" book on yacht design explaining all the sailor needs to become an amateur designer. *American Small Sailing Craft* is a survey of the history and design of the small craft that were the inspiration for many early cruising yachts. See also his *Boatbuilding*.

Coles, Adlard, *Heavy Weather Sailing*, John De Graff, Inc., Tuckhoe, NY, 1968.

A superlative treatise on sailing in heavy conditions in the form of a series of case studies. Many editions and the later ones tend to have more material.

Curry, Manfred, *Yacht Racing, The Aerodynamics of Sails and Racing Tactics*, Charles Scribner's Sons, New York, 1933. First published 1928.

Manfred Curry was the first to apply the new science of aerodynamics to the art of yacht design. He promoted the high-aspect rig, full-length sail battens and studied the effect of headsail overlap. He was also a master racing tactician. His work is a necessity for the successful racing skipper.

Hasler, H. G., and McLeod, J. K., *Practical Junk Rig*, Adlard Coles, Southampton, 1988

A thorough discussion of the Chinese rig and how it may be adapted for small oceangoing yachts.

Herreshoff, L. Francis, *The Common Sense of Yacht Design*, The Rudder Publishing Co., New York, 1948.

Anyone interested in cruising yachts, boatbuilding, or yacht design must read this book. It was first published as a monthly series in *Rudder Magazine*.

———. *Sensible Cruising Designs*, International Marine Publishing Company, Camden, ME, 1973.

A collection of Herreshoff's cruising designs with both building instructions and commentary on how to use them.

Illingworth, J. H., *Offshore*, Adlard Coles Ltd., Southampton, England, 1949.

———. *Further Offshore*, sixth edition, completely revised, One Design and *Offshore Yachtsman Magazine*, Chicago, 1969.

Illingworth was an English ocean racer who addressed the problems of ocean racing using science and engineering. Although compromised by outdated rating rules, his books are a valuable source of experience.

Kay, H. F., *The Science of Yachts, Wind & Water*, John de Graff, Inc., Tuckahoe, NY, 1971.

Kemp, Dixon, *Yacht Architecture, A Treatise*, Horace Cox, London, 1891.

A massive and early treatise, one of the first to deal with small sailing boats.

Kinney, Francis S., *Skene's Elements of Yacht Design*, Dodd, Mead & Company, New York, 1973.

Francis Kinney worked for many years at Sparkman and Stephens. This book is his expansion of Norman Skene's work below. It is in fact a new work but endeavors (successfully) to do the same job as Skene's book, but forty-five years later.

Marchaj, C. A., *Sailing Theory and Practice*, Dodd, Mead & Company, New York, 1962, and revised edition 1984.

———. *Aero-hydrodynamics of Sailing*, Dodd, Mead & Company, New York, 1979.

———. *Seaworthiness: The Forgotten Factor*, International Marine Publishing Company, Camden, ME, 1986.

———. *Sail Performance*, International Marine Publishing Company, Camden, ME, 1990.

Marchaj's work remains the authority on the science of sailing. He covers the field so well, his books have yet to be surpassed. *Seaworthiness* is especially important as it discusses how the modern ocean racer developed beyond excellence into the decadence of unseaworthiness.

Phillips-Birt, D., *The Naval Architecture of Small Craft*, Hutchinson, London, 1957.

A classic English text by a man with a good eye.

Saunders, Harold E., *Hydrodynamics in Ship Design*, volumes I and II, The Society of Naval Architects and Marine Engineers, New York, 1957.

A massive work, beautifully written, covering all aspects of displacement vessels moving through water. A necessity for the professional and a reference for the sailor.

Skene, Norman L., *Elements of Yacht Design*, Kennedy Bros., Inc., New York, 1927. See Kinney above.

Norman Skene worked for most of the early 20th century American designers and brought the necessary science to their art. His book became the text upon which they all depended. When I mentioned my interest in yacht design to William Tripp, Jr., at the 1959 New York Boat Show, his only comment was, "Read Skene."

Street, Donald, *The Ocean Sailing Yacht*, W. W. Norton & Company, Inc., New York, 1973.

A brilliant book written by an extraordinary sailor on how to outfit and sail an oceangoing sailboat. Written just before the world went gadget crazy.

BOATBUILDING

Boatbuilding books are not so common, probably because being a commercial enterprise building details are trade secrets closely kept.

Chapelle, Howard I., *Boatbuilding*, W. W. Norton, New York, 1941, 1985.

In making his survey of surviving small craft along the American coasts in the 1930's, Chapelle became expert in their traditional construction. This book gives a detailed description of how it works and how you can do it.

Gougeon Brothers, Inc., *Gougeon Brothers on Boatbuilding*, Bay City, MI, 1979.

The three Gougeon brothers began using epoxy resins in the 1960's to replace metal fasteners for boatbuilding and developed a business for creating epoxy products for boatbuilders. This book promotes their products by explaining how to use them for cold-molding and other boatbuilding tasks. It is a primer for epoxy-bonded boatbuilding and, after forty years, still a useful handbook.

Klingel, Gilbert C., *Boatbuilding with Steel*, International Marine Publishing Company, Camden, ME, 1973.

An introduction to steel boatbuilding by the owner of a building yard.

Watts, C. J., *Practical Yacht Construction*, Adlard Coles, London, 1947.

A short treatise on how Camper and Nicholson built wood sailboats in the 1950's, written by their chief draftsman.

REFERENCES

Dexter, Steven C., *Handbook of Oceanographic Engineering Materials*, Robert E. Kreiger Publishing Company, Malabar, FL, 1979.

Dexter's is an extremely valuable handbook of information about structural materials for use in and near seawater. It gives mechanical constants, corrosion rates, availability, and working characteristics. Wood gets 6 of the 314 pages! Hard to find. Worth the effort.

Encyclopedia of Wood, Sterling, N Y, 1989.

This is a reprint of *Wood Handbook*, originally published by U. S. Forest Products Laboratory. Basic and useful information focusing on the commercial uses of American woods with some information on commonly imported species.

Gibbs and Cox, *Marine Design Manual for Fiberglass Reinforced Plastics*, McGraw Hill, New York, 1960.

Early in the fiberglass age, Owens Corning retained Gibbs and Cox, the marine engineering firm, to establish standards and strength properties of the new material. It remains the authority for polyester-based, glass-reinforced laminates.

Handbook of Hardwoods, Her Majesty's Stationery Office, London, 1972.

An English handbook giving the principal physical facts about many species of wood worldwide.

Timber Design and Construction Handbook, Timber Engineering Company, McGraw Hill, New York, 1956.

Industry association structural handbook.

ABOUT THE AUTHOR

GOD CREATED ME A SAILBOAT NUT. I was born in 1942 and raised in Tennessee, but I spent summers in Nantucket, where my passion for sailing manifested itself. My father taught me about the sea and sailing aboard his boats, and I passed the Tennessee months drawing, building, and reading about boats in Pop's shop and library.

I was sent to school at Exeter, where besides reading, writing, and arithmetic, I was taught celestial navigation. After mathematics at Harvard I followed Nathanael Herreshoff and Olin Stephens to MIT, from which I graduated first in class at the Architectural School.

In 1977, my brother Edward and I began America's first boat shop doing cold-molding on a production basis. Our product was a replica of Nat Herreshoff's personal boat, *ALERION*. At Sanford Boat, between 1978 and 1982, we produced 21 *Alerion Class Sloops* and oversaw the construction of 8 more in later years.

In 1981, my friend Rick Wood and I transformed Blue Bahia Boatyard in Richmond, California, into a major Bay boatyard known as Sanford-Wood Marine. Sanford-Wood was a repair yard for yachts and fish boats, but we also built *FANCY* and a couple of steel boats, including Bernard Moitessier's last boat, *TOMATA*. Sanford-Wood continues today as Keefe Kaplan Marine.

I began ocean sailing in 1956 with a Bahamian cruise on *MALABAR X*. Ocean racing began in 1960, and I skippered offshore for the first time in the 1963 Marblehead-Halifax race. I made an effort, with some success, to indulge myself with a month a year on the ocean. I have been privileged to sail many miles, aboard 29 different boats, usually as skipper or navigator, around the edges of the Atlantic and across it, from the Bahamas to the Norwegian Sea to the Aegean, including 14 voyages to Bermuda (7 racing), 8 roundtrips from Nantucket to the Bahamas, four to the Maritimes, and 4 transatlantic passages, 3 in *IMPALA*, my lovely 57 foot S&S yawl.

Wooden Boats for Blue Water Sailors is my second book. I wrote the pilot for Nantucket waters, *Sailing around Nantucket*, in 2015.

CREDITS

With the exceptions listed below, the photos in this book are mine. Likewise, most of the drawings and diagrams, are my originals. Further I have added editorial comments or modified for clarity illustrations 1-15, 7-1, 10-3, 10-6, and 10-7.

But some of the finest work are scans of drawings and photos from books in my library. A few of the best are from original work in museums. These I note below.

Where possible, I have obtained permission from the owners of the material. But in several cases both the original creators and their publishers are no longer with us, leaving me only with the option of crediting them for their work and encouraging my readers to study their other work as well. Without exception their accomplishments are exquisite.

	Cover photo	Jim Powers courtesy of Nantucket's *Inquirer and Mirror*.
1-1	*NORTHERN LIGHT*	New Bedford Whaling Museum.
1-2	*SPRAY*	New Bedford Whaling Museum.
1-3	*PRESTO*	Scanned from Gilpin, *The Good Little Ship*, Vince Gilpin, facing p. 8, p. 19, p. 39. Copyright owner sought but not found.
1-4	*ISLANDER*	Scanned from Pidgeon, *Around the World Single-Handed*, facing pp. 8-9, p. 13, facing p. 161. Copyright owner sought but not found.
1-5	*ISLANDER*	Scanned from *National Geographic Magazine*, February 1928, p. 148.
1-6	*ALICE*	Scanned from Howard, *The Yacht Alice*, frontispiece. Copyright owner sought but not found.

1-7	*FORE & AFT*	Scanned from Atkin, *Of Yachts and Men*, p. 91, p. 93, p. 95, p. 109. Copyright owner sought but not found.
1-8	*SVAAP*	Mystic Seaport Museum, Rosenfeld Collection.
1-9	*MALABAR X*	Mystic Seaport Museum, Rosenfeld Collection.
1-10	*LEGH II*	Scanned from Dumas, *Alone through the Roaring Forties*, facing p. 128, frontispiece, p. ix. Copyright owner sought but not found.
1-11	*VARUA*	Scanned from Robinson, *To the Great Southern Sea*, facing p. 136. Copyright owner sought but not found.
1-12	*LONE GULL*	Scanned from Griffiths, *Dream Ships*, p. 118. Copyright owner sought but not found.
1-13	*TREKKA*	Maritime Museum of British Columbia.
1-14	*AVENGER*	Uffa Fox On-Line, compliments Mike Dixon.
1-15	*Flying Fifteen*	Scanned from Fox, *Sailing Boats*, p. 90, compliments Mike Dixon.
1-16	*HOOT MON*	Mystic Seaport Museum, Rosenfeld Collection.
3-2	*ALERION [III]*	Mystic Seaport Museum, Rosenfeld Collection.
7-1	*R boat, YANKEE*	Scanned from Herreshoff, *The Commonsense of Yacht Design*, p. 63, with permission from Mystic Seaport Museum.
10-3	Stems	Scanned from Watts, *Practical Yacht Construction*, p. 17, p. 20. Copyright owner sought but not found. Editorial comments by author.
10-6	Floors	Scanned from Chapelle, *Boatbuilding*, p. 202. Copyright owner sought but not found. Editorial comments by author.
10-7	Keel/floor	Scanned from Herreshoff, *The Commonsense of Yacht Design*, p. 73, with permission from Mystic Seaport Museum.
10-17	Bootstripe	Scanned from Fox, *Sail and Power*, p. 95, compliments Mike Dixon.
15-1	*CURLEW*	Scanned from *Antarctic Oasis*, pp. 106-107. Copyright owner sought but not found.
15-2	*IMPALA*	Jim Powers, courtesy of *The Inquirer & Mirror*, Nantucket, MA.

CREDITS

The advance readers of this book, listed below, offered suggestions, corrections and additions. They are old friends, professional acquaintances and *IMPALA* sailors. I introduce them and thank them, as follows:

George Bahen, Stuart, Florida. Chesapeake Bay sailor, voyager, explorer of French canals, delivery skipper and long-time *IMPALA* sailor.

Maynard Bray, Brooklin, Maine. Master boat builder, sailor, WoodenBoat contributor, author and Herreshoff specialist.

Halsey Herreshoff, Bristol, Rhode Island. Yacht designer, America's Cup sailor, voyager, and Captain Nat's grandson.

Eric Holch, Nantucket, Massachusetts. Graphic artist, sailor, former Commodore and long-time IMPALA sailor.

Laurie McGowan, Monchelle, Nova Scotia. Naval architect, very inventive designer and CAD technician.

Bill Page, Cushing, Maine. Yacht broker, boat builder, voyager, and classic marine hardware collector.

Matt Smith, Barrington, Rhode Island. Naval architect, protégé of Ted Hood, brilliant designer, and expert CAD loftsman.

Barney Sandeman, Poole, England. Classic yacht broker, sailor, and owner of the beautiful Sparkman and Stephens yawl, *LAUGHING GULL*.

Roger Taylor, aboard *WATERLILY*. Writer, yachting historian, voyager and publisher of nautical books.

Michael Walker, Shreveport, Louisiana. Michael skippered *ASPARA* on her voyage mentioned p.34, He is and has been a sailor, ship manager, marine engineer, and long-time *IMPALA* sailor.

Steve White, Brooklin, Maine. Brooklin Boat Yard's president, repairer of and sailor aboard, *IMPALA*, foremost builder of cold-molded sailing yachts in the world.

www.ingramcontent.com/pod-product-compliance
Lightning Source LLC
Chambersburg PA
CBHW061224150426
42811CB00057BB/1271